World Disasters Report

1997

International Federation
of Red Cross and Red Crescent Societies

Oxford University Press
1997

Acknowledgements

The World Disasters Report 1997 was edited by Nick Cater and Peter Walker.

Design and layout: Nikki Meith, *Maximedia.* Production manager: Sue Pfiffner.

Principal contributors: Chapter 1, John Sparrow; Box 1.4, Stephen Davey, IFRC; Chapter 2, Charles Dobbie; Chapter 3, Linda Stoddart, IFRC; Nick Cater; Box 3.1, Greg Swarts; Box 3.2, UNDHA; Box 3.3, John Black, IFRC; Chapter 4, Eric Noji, Centers for Disease Control; Chapter 5, John Borton, Joanna Macrae, Moira Reddick, Overseas Development Institute; Box 5.1, Jón Valfells, IFRC; Chapter 6, Andrew Hall, IFRC; Nick Cater; Chapter 7, Denis McClean, IFRC; Chapter 8, Thorir Gudmundsson, IFRC; Chapter 9, Joachim Kreysler, IFRC; Box 9.4, Nick Cater; Chapter 10, Centre for Research on the Epidemiology of Disasters; Department of Peace and Conflict Research, Uppsala University; US Committee for Refugees; Chapter 11, Peter Walker, IFRC; Box 11.3, Nick Cater; Chapter 12, IFRC; Chapter 13, IFRC. Thanks to all those who assisted contributors during travel and research.

Photographs: Cover, Howard J. Davies/ International Federation. Chapter 1, Heine Pedersen/ International Federation. Chapter 2, Howard J. Davies/International Federation. Chapter 3, Chris Black/International Federation. Chapter 4, Heine Pedersen/International Federation. Chapter 5, Oleg Litvin/International Federation. Chapter 6, Heine Pedersen/ International Federation. Chapter 7, © Kevin West. Chapter 8, Thorir Gudmundsson/International Federation. Chapter 9, Gary Carlton/International Federation. Chapter 10, Heine Pedersen/International Federation. Chapter 11, Howard J. Davies/International Federation. Chapter 12, Damien Personnaz/International Federation. Chapter 13, Lars Schwetje/ International Federation.

Contact details

International Federation of Red Cross and Red Crescent Societies
17, chemin des Crêts, P.O. Box 372,
1211 Geneva 19, Switzerland
Tel: (41)(22) 730 4222
Fax: (41)(22) 733 0395
E-mail: secretariat@ifrc.org

Editing

Nick Cater
Words & Pictures
Tudor St Anthony, Muchelney
Langport, Somerset TA10 0DL, UK
Tel: (44)(1458) 251 727
Fax: (44)(1458) 251 749
E-mail: cater@ifrc.org

Contents

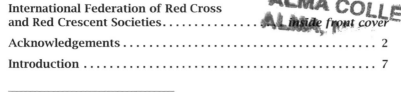

Section One Key Issues

Section Two Methodologies

Section Five Red Cross and Red Crescent

Faith, hope and standards

Three key sets of organizations today provide the support services for the poor and vulnerable people of the world: states, with their welfare systems; companies, with pay-as-you-need services; and civil society, with its plethora of agencies, religious organizations, self-help groups and political associations. People once put their faith in the state, hoped for jobs or cheap services from companies, and if all else failed took the charity of civil society.

Faith, hope and charity are changing. The ability of states to directly provide for vulnerable people is waning, nationally and internationally. Companies can pick up some of the load by running welfare services for a profit. Civil society is increasingly expected to provide for the needs of vulnerable people. Civil society in developed countries is experiencing growth rates in turnover and jobs that outstrip the rest of their economies. International humanitarian assistance – most of which is eventually delivered by the International Red Cross and Red Crescent Movement and non-governmental organizations (NGOs) – has been booming.

Agencies no longer fill gaps; they are now at the centre of the challenge of caring for and empowering the most vulnerable. This new positioning gives humanitarian agencies a special responsibility to understand the business they work in. Agencies have to be clear about who they assist, how and why. Unlike companies, agencies do not have profit or loss, but multiple "bottom lines" and diverse "customers" in disasters.

Disaster victims have an important stake in disaster response. Their critical concern is how the split is made between those who receive assistance and those who do not. Once victims are beneficiaries, they become – in this market-like model – the end-point consumers, but with almost no say in the quantity and quality of the service they receive. Further developing the *Code of Conduct for the International Red Cross and Red Crescent Movement and NGOs in Disaster Relief* will help protect the rights of disaster beneficiaries; work underway to set down universal quality standards for disaster assistance will help even more.

> **Agencies are now at the centre of the challenge of caring for and empowering the most vulnerable.**

A generation ago, most international disaster assistance went from government to government, sometimes via the United Nations. Not so today. Most end-point providers of international assistance are independent, private, not-for-profit, non-governmental agencies. The Red Cross and Red Crescent, plus the familiar names of big Northern charities, deliver the vast bulk of international relief, alongside hundreds of smaller agencies. What stake do they have in disasters?

These NGOs – whose changing role is a key theme of the *World Disasters Report 1997* – are like firemen, hoping there will be no fire, but knowing that their survival depends on having fires to put out. The NGO is a complex entity. It must balance its freedom and independence against the power to act that comes with large govern-

Introduction

ment or private donations, and balance the need to cooperate with others operationally against the urge to compete for funds and media exposure. Are NGOs still the idealistic organizations most once were, or have they been forced to compromise their values?

Suffering people, idealistic organizations and, finally, new forces seeking new roles. Peacekeeping and peacemaking are growing markets for Northern military forces and many are keen to see swords beaten into the ploughshares of humanitarian action. But is their stake in disasters compatible with that of disaster victims and humanitarian agencies? Can foreign military ever be truly humanitarian, neutral and independent in practice?

> *...ethically-based life and death decisions require the guidance of agreed standards.*

In most markets, all this competition would be a positive force, increasing value for consumers by raising standards, cutting prices or both. But humanitarian assistance is not that simple: its ethically-based life and death decisions require the guidance of agreed standards; its final consumers have limited political or economic power. Competition will only improve value if funders become more discerning, and like-minded agencies see the advantage in setting high standards for themselves.

The choice of higher or lower standards lies with operational agencies and those who fund them. As yet, the end-point consumer – the disaster victim – has little choice.

George Weber
Secretary General

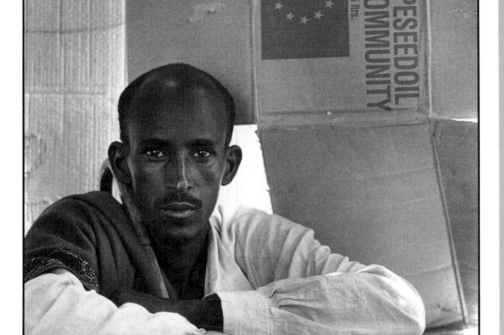

Chapter

Changing NGOs and the crisis of confidence

The hottest places in hell are reserved for those who stay neutral in times of moral crisis, concluded the visionary Italian, Dante, in *The Divine Comedy* almost seven centuries ago. Exiled for his views, he also understood that independence can be hellish. In today's complex world of conflict and disaster, aid agencies are rediscovering suffering amid many challenges – financial, political, ethical – that have brought a crisis of confidence.

At root are issues far larger than disaster response or development aid: the future of the nation state, and where the state stops and non-state institutions – charities, companies, community groups and more – take over.

In a single ideology world, many governments are disengaging from poorer countries where they see little political or trade interests. Instead they welcome a growing flow of private investment into the South, even though that usually bypasses the countries and people most in need. Retreating governments – failed or frayed Southern states, cash-strapped Northern administrations, market-orientated Eastern economic "tigers" – are leaving more for others to do, usually with inadequate resources, including international welfare.

> **...the battle for funds, media profile and market share appears unstoppable.**

It is not just that governments want to buy a slice of agencies and their values; governments themselves do not believe governments can deliver. Those being "bought" are asking hard questions about how far they are losing independence, covering up government failure, doing too much on the cheap or doing more harm than good. Some find it disturbing to be characterized as part of the global trends of deregulation and privatization, accused of undermining Southern states by providing foreign-funded alternatives and eroding legitimacy in ways that are neocolonialist, if not subversive.

Forgetting politics, for many the problem is cold hard cash, or the lack of it. Official aid is falling, even cash for high-profile disasters may have peaked and the collecting tin of public giving faces more competition and recession. Big cuts have begun.

Meanwhile, there is competition at the sharp end in disasters, from military forces looking for work to companies looking for profits, and the new diversity of groups – human rights, conflict resolution, gender, even environmental – found in greater numbers at each emergency.

Contradictions abound. All aid agencies talk of working themselves out of a job, but despite the massive growth of indigenous Southern development and disaster groups, the battle for funds, media profile and market share appears unstoppable.

Agency doubters are matched by outside critics as Cold War certainties become new complexities. What are aid agencies – partners, prophets, palliative or prescription – and have they lost their way? To ask seems profane; much has been done. In 30 years, malnutrition has fallen 30 per cent, infant mortality has halved, the rural poor with safe water rose from 10 to 60 per cent. Michael Edwards, ex-head of information and research for the Save the Children Fund-UK, does not accept there is a crisis but finds agencies under great pressure: "We are at a very major crossroads."

Agencies were born of disasters, and where the Red Cross and Red Crescent once toiled alone – started by Henry Dunant, shocked by suffering at the battle of Solferino – many dynamic agencies have emerged. Outraged by a post-war Allied blockade causing child starvation in Austria, Eglantyne Jebb founded the Save the Children Fund (SCF) in Britain in 1919 with a declaration that evolved into the United Nations (UN) Convention on the Rights of the Child. Launched in 1942 as the Oxford Committee for Famine Relief, Oxfam sought to aid the hungry of war-time Europe. CARE was close behind in North America, starting life as a food and clothing package service to post-war Europe and Asia.

Starving baby

By the 1960s, hardly slowed by the dreary label of non-governmental organization (NGO), agencies had spread worldwide. As crises passed, they stayed to tackle other needs, rebuild communities, or advocate just ones, filling gaps left by church or state, or jumping in where others feared to tread.

Public support opened a conduit for the conscience – and cash – of the comfortable, and Biafra set the tone. The Nigerian conflict put many thousands in peril, and gave birth to yet another major NGO, Médecins sans Frontières (MSF). An image of a starving baby imprinted itself on the psyche of the 1960s.

Today's multi-billion-dollar aid industry is a long way from Biafra's baby. Agencies put more money into Africa than the World Bank, mostly from Western governments. People-to-people, can-rattling "real aid" seems peripheral to the big business and media circus of today's NGOs. Disaster response went from zero to boom in 20 years. In 1971, 200 million US dollars of government aid – always far more than public donations – went on disasters; by 1994 Rwanda cost $1.4 billion as crisis spending reached $8 billion, but total official development aid had already peaked at $60 billion in 1993.

NGO numbers are still booming. In 1995, the Commission on Global Governance found 28,900 international NGOs. The Organisation for Economic Co-operation and Development (OECD) reported Northern development NGO numbers rising from 1,600 in 1980 to 2,970 in 1993. Growth may be even faster in the South. In 1994, the

---

Box 1.1 **Neutral and impartial – but are they really independent?**

Georgie Jakshich, 79, had not moved from his Sarajevo apartment for four months. He lay listening to shells demolish the building. One room of his apartment remained intact; daylight poured into the rest and snipers were 200 metres away. If they did not get him, the cold would. A glass of tea someone left froze before Georgie could drink it. The building had no heating, no electricity, no water supply; common problems after months of siege.

The Dutch government saw energy as vital. Shelling and sabotage had hit electricity; insufficient oil and coal were getting through. Cold homes were only one result. Water supply, bakeries and hospitals were all affected. Operations were performed under car headlights hooked to batteries. Natural gas still came down a pipeline from Russia into a network that the gas-rich Netherlands had helped develop, but its systems needed electricity. The plan: supply gas-fired electricity generators, and restore gas cooking and heating to homes and institutions.

The Netherlands Red Cross received 11 million guilders ($6.5 million) from the Dutch government. In 18 months, with ICRC support, 141 generators were flown in and installed, 40,000 apartments were restored to hot-water central heating, thousands of new heaters and cookers were working, and many solid-fuel stoves had been converted to gas. Bakery production soared, gas generators helped pump 800,000 litres of water a day. Gas was the saviour of Sarajevo, said Dr Bakir Nakas, Director of a re-energized state hospital.

As it took government cash to meet the gas challenge, how independent is the independent Red Cross and Red Crescent? International spending by the Netherlands Red Cross is up seven-fold in a decade, reaching 70 million guilders in 1995. Around 60 to 70 per cent of its funds for expansion into Africa,

Eastern Europe, the Middle East, Asia and former Yugoslavia came from the government and ECHO. That is not unusual. At a meeting of national Red Cross and Red Crescent societies on former Yugoslavia, only Austria and Iceland had sizeable reserves from the general public.

Jaap Timmer, Netherlands Red Cross head of international activities, is not worried. "Having generous partners doesn't compromise principle. Take Sarajevo. We did it on the understanding that all sides of the conflict were offered assistance. We talked with the Serbs as well as the Bosnian Moslems. Our ministry agreed. The Red Cross is neutral, impartial and transparent. That's why donors come to us. That and our experience. That's our comparative advantage."

Timmer concedes government or ECHO concern for certain regions and problems influences humanitarian direction, but argues partnership is a two-way street. "One day it's the donor asking for help, the next day the Red Cross asking for funding. Where concern is common and ethical, what's the problem?" Besides, government spending reflects public opinion which humanitarians should be influencing. "Advocacy from our side is paramount."

But he says: "We will never tender for contracts. Humanitarian help is about quality, impact and working methods. Efficiency alone is not enough. You need ethical principles, something a commercial firm doesn't have."

He does not rule out the private sector, using firms widely as implementers – although kept away from policy – and welcoming sponsorship. KLM Royal Dutch Airlines funded relief after a Caribbean hurricane. Emphasis remains on widening sources. You need so many eggs, argues Timmer, and they will not be found in one basket.

Union of International Associations found worldwide more than 16,000 NGOs operating in three or more countries, and financed from more than one. The figure had more than doubled in three years.

Those are the ones recorded somewhere. Each crisis attracts new NGOs. The 1996 international evaluation of the Rwanda crisis could not find almost a third of the 170 agencies registered in the Great Lakes region. The destination of $120 million of the $1.4 billion was uncertain. Insiders are not surprised. In Somalia, local firms became NGOs overnight to get UN funds. In a humanitarian world without rules or a controlling body, anybody can be a relief agency.

> *The essence of the aid agency had always been values.*

The essence of the aid agency had always been values: raising its own funds to do its own thing. Could Jebb of SCF have confronted an Allied blockade if sponsored by Allied cash? But independence came under threat as government funding of NGOs became the norm. So much official aid has been available that purists among agencies have had to set limits for how much they will accept from governments or any one donor. These figures vary, but there are agencies taking 80 to 90 per cent of their budgets from governments. Nordic agencies, some of them totally state-funded, say this is no big deal, as politically-neutral Nordic governments do not politicize aid. Agencies consider themselves unhindered. Elsewhere, independence is a sensitive topic.

Continuing trend

The fear is for values when politics – buying influence, drawing lines on maps – can define need and make funding conditional. Contracts are replacing old no-strings grants for agencies. If agencies bid for contracts and donors set terms, where is independence?

The trend may continue. Although the American NGO coalition InterAction says that faster-rising public giving is making agencies less dependent on government than a decade ago, 400 US NGOs took a third of their money from government in 1995. The US Agency for International Development (USAID) is pledged to put 40 per cent of its aid through NGOs by the year 2000. It climbed from 13 per cent in 1992 to 28 per cent in 1995. UK aid agencies already rely on government for more than 35 per cent of their budgets.

Some see a conspiracy to hijack free-wheeling humanitarians, as Western governments have increasingly turned to NGOs to escape from tough problems. Chaos in Cambodia from 1979 to 1982 was a watershed. Ironically, an Oxfam-led NGO consortium showed governments the way. Politics paralysed the ability or willingness of Western governments to assist. The West condemned Viet Nam's occupation to overthrow the Khmer Rouge, and the Khmer Rouge kept Cambodia's UN seat. The Vietnamese-backed government insisted aid come through Phnom Penh; the Khmer Rouge wanted it to go through the Thai border areas to which it had retreated. The International Committee of the Red Cross (ICRC) and UN agencies responded with a joint mission, but it was the 35-member NGO consortium that caught Western governments' eyes. The consortium spent $45 million, a scale of operation normally reserved for government. NGOs were no longer a sideshow.

Says Brian Walker, Oxfam director from 1974 to 1983, "It was efficient, it delivered the goods, and we operated entirely on a humanitarian basis regardless of political restraints. The Cambodians, for example, were self-sufficient in seed production inside a year." Governments saw the potential. Walker added: "It might have been the opposite of what they wanted. They wanted to break Cambodia and Viet Nam as part of the Cold War process. We helped stabilize the country and gave a new and different view of what Viet Nam was doing. Governments realized we had the capacity and began to see they could reduce their costs...by pushing money through NGOs."

For donors reluctant to give aid to corrupt or inefficient states, NGOs were a controllable alternative, allowing donors to cut overseas commitments, cost-effectively privatizing aid, but still get performance. Government-to-government aid fell, NGO funding climbed.

Today Alex de Waal of African Rights observes falling Western strategic and commercial interests in poor countries, and governments trying to avoid bad publicity for inaction if crises hit the TV news. NGO funding is "high profile, flexible, short term, and has little accountability."

NGOs are frontline troops for governments which prefer humanitarian help to political solutions. Walker laments it. He limited official funding of Oxfam to 10 per cent, kept government at arm's length and never had any interference. "That seems to have gone by the board for everyone." He describes NGOs as a thermometer for the health of a nation state. If they betray their past and become tools of government, all of society is diminished. Better to have less cash and talk from experience with integrity, he argues, than get millions more and echo governments.

Funding with strings is the price of being part of conventional wisdom, says Rudy von Bernuth, Executive Director of the Geneva-based International Council of Voluntary Agencies (ICVA). Peripheral, independent life gave NGOs moral superiority. As others back the NGO approach, they acquire new stockholders. "I don't think that's

---

Box 1.2 **Coordination, cooperation, liaison – and getting it together**

The Great Lakes scramble left some asking whether agencies will ever get their act together. The exception was Tanzania, where teamwork took the place of the agency overlap and competition in Zaire, Burundi and Rwanda. ECHO and USAID put all funding through UNHCR, which also had full authority from the Tanzanian government to approve agencies. It rejected 40; successful agencies were unanimous that cooperation had never been better. Coordination triumphed.

Donors were delighted. Sean Greenaway, of ECHO's department of strategic planning and policy analysis, says it was a daring solution that avoided the risky alternative of funding agencies separately. ECHO does not consider it a definitive answer. "But it did work there and, given similar conditions, we would not hesitate to repeat the formula."

Looking at the broader picture, Martin Griffiths of UNDHA remains a realist in lobbying for closer UN-NGO relationships. He finds that "many UN people see NGOs as an irritant when it comes to substantial, serious and grown-up discussions. The real gap lies in NGO involvement at policy level." UNDHA's inter-agency standing committee, where NGOs and UN are equal players in policy debate, has no executive authority. Griffiths says: "It is not full-blooded but at least it's recognition that there should be involvement at policy level."

In this trend, he advocates three steps. NGOs must translate experience into policy positions. A systematic UN-NGO relationship must decide on policy as well as operational issues. Agency personnel exchanges through secondments – now rare – will help remove an attitude problem. To move the UN, NGOs will have to persuade governments that a policy dialogue is needed. He adds that donors themselves must improve coordination.

As the International Federation and NGOs seek universal standards, the 150 members of the US NGO coalition InterAction have signed a self-enforcing field protocol to cooperate.

Julia Taft, InterAction's President, says: "Cooperation among disaster responders has become more imperative. Victims cannot be left untreated while NGOs shift resources to anticipate CNN coverage. Precious NGO funds should not be consumed in bidding battles for local resources. Incompatible communication nets should not expose NGOs to security risks."

ECHO is pushing field coordination structures and information sharing, swapping operational information with EU states, keeping close contact with USAID, and talking with non-EU donors. Its operational statistics are on the Internet, as will soon be those of its member states. It also supports UNDHA's ReliefWeb.

---

a problem. If NGOs work it right, they are using public assistance to promote their way of doing things." NGOs must know their agenda.

As the aid boom levels off, agencies are cutting back. Oxfam-UK, with a GB£ 5.6 million shortfall, announced in 1996 a 20 per cent reduction of staff over two or three years, and began reviewing overseas programmes. SCF-UK is cutting £9.5 million over three years. Goal of Ireland, £2 million short, said it would phase out of the Great Lakes region and other areas, losing up to one-third of its volunteers. Some Canadian agencies have closed.

Von Bernuth sees more restructuring ahead, a "winnowing out". There is talk of mergers. Less public giving is part of the problem. Goal director John O'Shea blamed a fall in awareness caused by a lack of disasters on television. Oxfam reported lower donations.

Official relief funding looks set to stagnate or fall. The European Community Humanitarian Office (ECHO) reflects that. From 368 million Ecu in 1992, its 1993

Box 1.3 Learning how not to aid the war effort

Does aid do any good or, more fundamentally, can it actually do harm? On both sides of the Atlantic, researchers are arguing that aid can exacerbate conflict and undermine states.

In the booklet *Do No Harm*, Mary Anderson of the US-based Collaborative for Development Action, catalogues how aid goes wrong and suggests approaches to avoid the negative and pursue the positive.

Aid, she argues, can deepen and prolong conflicts whose effects it tries to address. "Even as aid helps, it also harms. The act of mercy, once believed to be simple and pure, is complicated and politically implicated in the conditions of today's world."

Relief goods brought to conflict zones can be taxed, stolen, diverted or manipulated by warring parties to support their efforts or enhance their power over civilian populations.

Aid systems often carry implicit ethical messages about power and violence. Agencies which negotiate with military leaders to get access to civilians, or which hire armed guards to protect goods, appear to accept the right of arms to determine who gets assisted.

Also warning that "bulk aid" can fuel war is Professor Paul Richards of the Joint War-Peace Research Programme of University College London and Wageningen University in the Netherlands. Studies of West African conflicts have led him to conclude that "although bulk supplies may often prove unavoidable, it is important where possible to emphasize `smart' (knowledge-intensive) forms of relief", such as using radio to transmit survival information and stimulate cultural processes.

Even in peace, there is concern about whether aid undermines states, what some call neo-colonial subversion. Ironically, if NGOs successfully provide welfare and social services, they may usurp the role of government, diminish its perceived legitimacy and encourage instability.

Anderson says Southern government officials lament this. "It's a critical issue that holds them back from being able to establish the kind of democratic government they want to. Ethiopia, in particular, wants to handle its own emergencies. It says it's ready. Western governments, meanwhile, are pouring the money through NGOs."

One highly-placed Mozambican figure demanded: "What can we do about the NGOs who are running our country? We are trying to take responsibility but we can't get the resources. Even if we could, we could never compete with the NGOs." Anderson cautions: "It is their country. It is their crisis. We proudly say we are trying to work ourselves out of a job. I think we don't mean it. NGOs are worried about their own longevity so they are hanging on to what they have got."

Anderson is not so concerned about small NGOs that arrive for one disaster and are not seen again. "The real harm is done by huge resources coming in and reinforcing dependency and conflict." Growing financial restraints may force action over negative impacts.

"It's almost naive to say so, but I do trust the humanitarian impulse. We have to respond, nurture and appreciate it. If we end up getting it wrong sometimes, the point is not to shout `stop'. It's to say: `Okay, how do we learn and get it right?'"

spending was 605 million, 764 million in 1994, 692 million in 1995, and was likely to fall again in 1996. Even Japan – the biggest bilateral donor – felt pressure on public expenditure.

A study of six British NGOs shows how they all grew fast on official aid, and all are now vulnerable. Its author, Tasneem Mowjee, of the London School of Economics, says new agencies ask for government funds because they cannot compete with the big players. Smaller agencies are the most dependent but argue that multiple sources reduce the risks. "There are signs that some agencies are about agencies. The talk is of growth. A cause comes next. I wondered why some are in business. When you ask what they're about, you get rhetoric. The word `beneficiary' doesn't always come up."

> *Everyone is following the money.*

A dependent, cash-strapped agency is not going where there are no funds. What of those who suffer in lands unpopular with Western governments?

In former Yugoslavia, mass repatriation of Bosnian refugees seems about to begin, despite warnings from the UN High Commissioner for Refugees (UNHCR) and other agencies that it is dangerously premature. Refugee host nations wanted a lead from Germany, which plans to repatriate 350,000 people. Positive concern and fears of negative media will mean plenty of repatriation funding. Despite the needs of the dislocated and impoverished in Serbia and Croatia, NGOs were relocating to Bosnia to get a share. Money and politics were the driving forces, not the welfare of conflict victims.

Three million Serbians – almost one-third – are in poverty, and 600,000 refugees face hard times, warns the International Federation of Red Cross and Red Crescent Societies. Peter Rees-Gildea, its regional desk officer, says the situation is "appalling, yet all anyone wants to do is build in Bosnia. Everyone is following the money."

Things could be worse. Amid funding problems, political sensitivities and a wary media, few tackled Iraq, where the Gulf War aftermath – especially UN sanctions – left hyperinflation, food and drugs shortages, serious malnutrition and collapsing health care. Three and a half million people were at risk, half a million of them children under five, reported the UN Children's Fund (UNICEF). The International Federation found appalling human suffering and unnecessary death, particularly among children.

Media support

If young Iraqis, dying from simple diseases for lack of drugs, were uninteresting, orphaned and lost Rwandan children, were – in the public-relations pimp talk that permeates agencies – "sexy". Those with mandates – UNICEF, SCF, ICRC and others – faced new young rivals in orphanages and tracing.

International Federation Secretary General, George Weber, still working hard to get interest in North Korean floods and hunger, says media support to raise money always makes an operation easier, "but we cannot give up on the others".

Going nowhere unpopular is a new force, the "Relief Inc." of private companies in trucking or telecoms are waiting to grab a share of official aid away from agencies by offering speed, efficiency, cost-effective performance and a willingness to take orders. Donors are listening; falling aid means a new focus on value for money.

Nan Borton, Director of the USAID Office of Foreign Disaster Assistance, does not see Relief Inc. as sinister, but as a potentially-useful partner for an increasingly-vulnerable global community. People in need should not be held hostage by labels. "Whether agency, NGO, bilateral or business, the proverbial bottom line is who can provide humanitarian services more effectively, sensitively and appropriately." She stresses: "Most important will be the standards that we set to demonstrate professionalism and the fulfilment of the humanitarian agenda."

NGOs must quantify their added value, says Martin Griffiths of the UN Department of Humanitarian Affairs (UNDHA). "If I were running a UN agency operation, I would certainly look at the private sector just as much as NGOs. It could be very good for beneficiaries."

Weber insists: "The private sector can provide efficient relief, but is it humanitarian? Think in terms of the value set. If you are driven by the profit margin, the need to keep your shareholders happy, are you going to provide aid in a humanitarian fashion, neutrally, impartially?"

Others think Relief Inc. will free them of drudgery, leaving the ethics, advocacy and judgement calls. Competitor or new partner, either view signals a shake-out. Old aid – once defined as living off left-inspired international solidarity, colonial hangovers, Judaeo-Christian guilt and post-war optimism – is on the way out.

> *Old aid is on the way out.*

But dependency on official aid or commercial competition is not the issue, says Nick Stockton, Emergencies Director of Oxfam-UK and Ireland. The question is rather: How much aid should go through non-governmental channels?

Stockton cites the view that aid can undermine political accountability by weakening social contracts between leaders and citizens. He forecasts more failed states unless a form of governance can be found that citizens perceive as legitimate. The best investments are social services, health care and education, but foreign donors will ease these away from the state. Losing legitimacy, states risk turmoil as they attempt to retain authority.

Not that Stockton rejects official aid. Taking government money gives agencies the chance to influence policy. The aloof are ignored. Official funds need balancing by private sources in adversity of support. "Cosying up" to governments, and reducing NGOs to contractors, detracts from a historic mission: organized socio-political expression.

A new age will not dawn without struggle, as shown last year when agencies and commercial suppliers met at WorldAid'96 in Geneva. An expo and conference to review products and services, and to discuss greater cooperation, it met a partial boycott from British NGOs. Michael Taylor, Director of Christian Aid-UK, told the conference he had been warned not to attend because he was in danger of endorsing what should be opposed. Firms should not profit from desperate need. But, he conceded, things were not quite so black and white.

Von Bernuth, whose ICVA led WorldAid, saw the boycott as peculiarly British. "The world in which we work has changed tremendously from the rather simple one of a decade ago. NGOs are in a crisis. Threatened at home by falling funds from the public, the last thing they need is competition, or somebody questioning their role at the centre of the humanitarian relief system. It's a false dichotomy. No one in the commercial sector remotely thinks that they should be setting the humanitarian agenda. That remains the prerogative of states, UN organizations and the NGOs, in some uneasy balance. Once the agenda has been set, more parties are coming to the table, and where they have comparative advantages we should be prepared to use them."

Going further than many, MSF is using companies as sponsors for some operations, and carries its funders' logos. "You have to be cautious," says Jean-Marie Kindermans,

Director of MSF International. "We wouldn't work with Shell in Nigeria. But there's no problem with corporate money, as long as you don't have to compromise." A hefty caveat, some would feel.

NGO networks are themselves big business. Ten CARE countries had a 1995 total income of $220 million, 60 per cent from official aid; MSF income from 19 countries was $265 million, up from $60 million ten years before, with 50 per cent – a self-imposed ceiling – from governments.

MSF has portrayed itself as the ultimate independent since 1971, when one-time Marxist activist Bernard Kouchner and dashing French doctors abandoned ICRC neutrality for "the right to inter-vene". Financial autonomy guarantees moral inde-pendence, it says, and in its main countries almost two million people donate regularly: 800,000 in France, 700,000 in the Netherlands, 400,000 in Belgium.

> *Seven or eight networks control 75 per cent of emergency-response capacity.*

But a global network also needs official aid, which carries risks. In 1995, the European Union (EU), critical of the new Rwandan government's policy, suspended aid to Kigali. MSF in Rwanda was partially ECHO-funded. MSF limits itself to 25 per cent from any one donor, and accepts nothing from those politically embroiled in a region. It avoided France in Rwanda, and the US in Somalia.

The latest network is Action by Churches Together (ACT), a web of 360 church groups and linked agencies that coordinates crisis response by the World Council of Churches (WCC) and the Lutheran World Federation. Myra Blyth, WCC Director of Sharing and Service, says: "Until ACT, Christians were sufficiently uncoordinated to compete among themselves for money." Now a single appeal can be out within 24 hours of an emergency.

NGO standby

Its Coordinator, Miriam Lutz, says: "Mistrust is a disease among NGOs. People lose sight of why we are there as they scramble for position, ownership, visibility. With ACT the objectives are clearer."

Seven or eight networks control 75 per cent of emergency-response capacity, according to ICVA. Their only certainty is change. Von Bernuth sees consolidation to come and points to the USAID introduction of the "indefinite quantity contract", putting NGOs on standby to respond on request within 72 hours to supply a service, from food to health.

He also expects donors will want NGOs to specialize. Oxfam for water and sanitation, say, CARE for logistics and delivery, Concern for nutritional rehabilitation, and MSF for acute health care. "It would make it easy for them to pick the right product off the shelf and plug it into the next disaster."

While Northern NGOs are consolidating, Southern NGOs are still growing fast. The UN Development Programme (UNDP) estimated in 1994 that there were 50,000 NGOs in the South. Tunisia had 5,200 in 1991, up from less than 1,900 in 1988; Bolivia in the 1980s went from 100 to 500; and Nepal had 1,000 new NGOs in the first three years of the 1990s. Bangladeshi NGOs are active in more than half the country's villages.

Large agencies are reaching millions: India's Working Women's Forum, Sri Lanka's SANASA, PROSHIKA in Bangladesh. The Bangladesh Rural Advancement Committee has as many staff, over 10,000, as CARE has worldwide. It operates in 15,000 villages and plans to increase its reach to three million people, empowering the poor through local groups and education programmes in 100,000 schools.

Southern NGOs are going international. Ethiopia-based Africa Humanitarian Action (AHA) offers relief and sustained development across the continent. The trigger was an apparent shift of Western NGO emphasis – chasing official funds – to Europe. AHA founder and President, Dr Dawit Zawde, says: "AHA is convinced the fundamental responsibility of alleviating the worsening crisis in Africa belongs to Africans."

In Asia, at least 5,000 NGOs helped draft the Filipino position for a recent Asia-Pacific Economic Cooperation Forum. People power makes some governments nervous. Indonesian activists accuse the authorities of intimidation and violence against

---

Box 1.4 The International Federation: an NGO or not?

Amid debates over the status, role and future of NGOs, how does one classify an organization that came into being over 75 years ago, before the term NGO gained common usage, and which claims to be both independent of government (an NGO?) and auxiliary to government (a para-statal?).

At first sight it seems that, since neither National Societies nor the International Federation are governmental organizations, they must be non-governmental, but such a short answer holds difficulties.

Dividing the world between the governmental and non-governmental excludes other entities, and yet the International Red Cross and Red Crescent Movement exists, and is unique. Although independent like NGOs, they enjoy close links with individual governments, and with international organizations made up of states.

If a government has recognized its national Red Cross or Red Crescent society in law as the only "auxiliary to the public authorities in the humanitarian field", that National Society enjoys a privileged relationship; independent of that government but with obligations to it, as well as a right to be heard and to be supported.

The meaning of the term "auxiliary" has changed over time, and tends to be interpreted differently from country to country. In most cases, it includes an element of being given a mandate by a government for certain tasks under specific conditions. This mandate is often referred to in the relevant legislation. The legislation is of great importance to the status of the National Society, its choice of activities and its ability to operate.

Through all those pieces of legislation, the International Federation, as the collective expression of National Societies, also has a privileged relationship with states. The functions of the International Federation are defined by states through the adoption of the Statutes of the Movement by the International Red Cross and Red Crescent Conference, where states, National Societies, the ICRC and the International Federation participate on an equal footing.

The term NGO is used by many organizations, in particular the UN and other intergovernmental bodies, as a criterion for representation at meetings, document distribution, and soliciting opinions. An NGO often has less access, less information and less weight.

Over the past few years the International Federation has increasingly been given a legal status similar to that of other international organizations, such as UN bodies and the ICRC. This provides it and National Societies with a platform and access to influence issues of importance.

The UN General Assembly has granted Observer Status to the International Federation, which means it participates in the work of the General Assembly and its subsidiary bodies on an ongoing basis, rather than being restricted to specific topics or individual meetings.

The International Federation also participates as an international organization in other UN fora; has a permanent invitation to the Interagency Standing Committee of the United Nations Humanitarian Agencies, a body trying to coordinate policies and practices within the field of humanitarian action; and has negotiated more than 30 status agreements with states, most recently with Switzerland, with others under negotiation.

As well as formal cooperation agreements with intergovernmental and regional organizations, such as the Organization of African Unity, the International Federation enjoys constructive negotiations over cooperation with regional development banks, such as the Inter-American Development Bank, and has the possibility to develop these with others, such as the European Bank for Reconstruction and Development.

5 million refugee

ethnic diversification
minorities

executive of the European Council
Refugees & Exiles
assessment visits
Bus service
2.2 million refugees

UNDP
clearing of minorities
areas.

twenty borders
asylum ffigure
army 5 of European refugees
demilitarization of Solodues

The Study of European Refugee

NGOs. The government intended to regulate foreign aid for NGOs, alleging members of some were involved in riots.

Most Southern NGOs have a high reliance on official aid and foreign sources. Foreign donors increased spending five-fold in Bangladesh over five years, and observers have noted donor-driven activity overriding indigenous priorities.

If cash goes South, the days of the quasi-colonial NGO are numbered. Von Bernuth says that for two decades the trend in sustainable development has been to work more through local partners. "Disaster response is coming late to this." He points to Turkey or India as models. "It's been a long time since India asked for any external assistance."

Major J. K. Michael, Director of India's Church's Auxiliary for Social Action (CASA), the relief and development arm of 24 Protestant and Ortho-dox churches, and an ACT member, would agree with that. As a decentralized disaster-response network, it tackles 60 to 70 disasters a year, from floods and fires to cyclones and earthquakes. The UN seeks its advice on disaster prepa-redness.

> *It's a mistake to believe in an organization's divine right to exist.*

How can the North help? Keep away. Or, as he politely puts it, "By being a genuine partner, assisting us...but understanding that decisions and planning should be left to the Indians." The Major knows that sending in teams, needed or not, means media, and media means money. "You get such a demand from agencies who want to fly over. We are so busy in calamities, there isn't time for guided tours. Friends are welcome. But give us a week until the system is moving."

Zenaida Delica, Executive Director of Citizens' Disaster Response Center (CDRC) in the Philippines, also sees a shift in the North-South relationship. The South needs plenty of help, she says, with education and advocacy in the North. Partners should not encroach on decision-making. Some act like doting parents, watching and intervening "for fear the partner will stumble. Often this results in misunderstan-ding."

Julia Taft, President of InterAction, agrees: "The time has come to fundamentally change the way we relate to the South, to no longer deal with affiliates but real partners. Ultimately, perhaps, we go out of business and leave them to it." InterAc-tion, she says, is pushing hard with the World Bank and UNDP to find and fund indigenous groups. "It would be wonderful if most of the resources went to the local NGOs and they subcontracted to the internationals for what they wanted. They would buy it from us. It would be very exciting, although we are a long way from a major breakthrough."

There is no consensus on the future. Michael Edwards says each NGO must resolve tensions between institutional and developmental imperatives, between what an agency thinks it must do to survive, and what it should do to honour its core values. "It's going to be diverse, unpredictable and messy. No one can predetermine the outcome but it requires us to stand firm for principles and values, and forcefully defend our independence."

He says the new agenda of the World Bank and other powerful institutions sees NGOs mainly as service providers in health, education, credit and relief, and rewards with funds those who accept this role. Proliferating Southern NGOs will displace their former Northern partners in many areas. On the surface, Northern NGOs have welcomed the rise of Southern NGOs. "Underneath, however, there are many ten-sions."

Northern NGOs must accept themselves as less important members of international alliances, and as real partners pursuing social change and common goals. "They now see themselves as world actors of the first rank. That's farcical; they are not that important. It's a mistake to believe in an organization's divine right to exist," he says. Agencies must change or become contractors until the money dries up. Many will not admit there is a problem: "We could be looking at a generational process. The quickest to respond will be the most successful."

Indigenous movement

Edwards says the crunch will come when donors fund Southern NGOs directly. Most donors' routes are still through Northern partners. "No one is hurting enough. When senior people at Britain's Overseas Development Administration or ECHO start saying these things, and back them up with action, change will occur very quickly."

He is echoed by Russ Kerr, Vice President of Relief and Rehabilitation for World Vision International, working in 65 countries: "Ten years down the road World Vision will be much more of a localized, indigenous movement. I see us in concert with smaller agencies, more involved with local groups rather than employing people, co-existing in alliances."

De Waal of African Rights believes agencies will remain naive about the future until they are accountable for the past. NGO meddling, he says, prolonged the Biafra war, refinanced the Khmer Rouge, and maintained Mengistu's grip on power in Ethiopia.

NGOs were never a solution to many of the problems they seek to address. "Humanitarians have never, for example, prevented famine, and if we are honest about it, humanitarianism is a palliative. Recognizing the limitations of NGOs is an essential step towards tackling the problem, which must be done at a political level." Agencies serve in a cynical way as a very important vehicle, not for solving anything but for appearing to. "They are a wonderful instrument for a US president or British prime minister, faced with popular outrage at starving people, to say: `I am doing something', while in fact doing nothing to resolve the problem."

His prescriptions are clear: Independent investigation for the structural lack of accountability and tighter regulation for the declining professionalism brought on by deregulation and privatization. He expects NGOs to go three ways. Some charlatans will chase publicity; others will lose their excitement and best people along with their research and public affairs departments to become public service contractors, like non-profit firms; a few will find a niche: small, radical and entrepreneurial, advocates and allies of neglected causes. Wasn't that where NGOs came from?

How do donors see the NGO future? Some themes are getting heavier emphasis: more pre-emptive action (preparedness, prevention), early response, meaningful regulation.

Accountable response

All donors stress more effective cooperation, more coordination. Jan Egeland, Norway's State Secretary for Foreign Affairs, says: "Building stronger bridges will become increasingly important, not only between NGOs, the UN, government agencies and the private sector, but also among NGOs themselves, and within the UN system."

Tomorrow's challenge for agencies is to maintain humanitarian values while making disaster response more efficient, effective and accountable.

Rwanda made clear that the time is overdue for universal standards of NGO behaviour. The *Code of Conduct for the International Red Cross and Red Crescent Movement and NGOs in Disaster Relief* (see Chapter 11) has been widely endorsed. That deals with ethics. Next will be practical standards for the quality of service disaster survivors can expect by right.

During WorldAid'96 a powerful group representing 95 per cent of front-line disaster-response agencies took a crucial first step. Networks in the Steering Committee for Humanitarian Response joined with InterAction to pool efforts on both sides of the Atlantic to create a single set of standards.

But calls are growing for external oversight, an independent ombudsman to arbitrate in disputes, to whom groups or individuals – including disaster victims themselves – can bring complaints.

Susan Blankhart, the Netherlands Foreign Ministry's Co-Director of Crisis Management and Humanitarian Assistance, says: "We have discussed this with other donors, and I would support an ombudsman within the UN. It cannot be someone from the NGO world. How can someone from one agency be in a position to pass judgement on those from another?"

> *You would not let self-regulation take care of any other sector.*

So, at least one question remained: If standards can be agreed, who will enforce them? For some agencies, regulation denies their very essence. But anything less risks being seen as a cover-up. Among agencies founded on values and integrity, will standards be standards, or statements of intent? One observer said: "You would not let self-regulation take care of any other sector."

Relief NGOs are a 20th-century phenomenon, following the rise and fall of colonialism, the growth of modern warfare and the expansion of trade. Conceived to meet an immediate need rather than aspiring to long-term goals, NGOs are under pressure to conform; to become agents of governments or intergovernmental bodies, to accept the pervasive market ethos or to adapt themselves to work alongside the political, economic and military manoeuvrings of nation states.

It is not possible to go back and wistfully re-establish maverick agency ancestors. The question is how can those committed to humanitarian ideals, to the alleviation of suffering and the practice of justice, adapt old institutions to still serve this ultimate purpose in a very changed world? Three trends are clear.

First, indigenous Southern NGOs will become far more important, both within their own countries and for external funders. They may well be the chosen conduits for future aid. But will they cope with the influx of funds and pressures any better than existing Northern NGOs?

Second, true North-South NGO alliances will be necessary. Lessons Northern NGOs are now learning need to be passed to indigenous organizations, while Northern NGOs rarely reach Southern grassroots opinion and aspirations. True alliances offer additionality, internationally and nationally. In Peru, for example, Lima squatter groups work with NGOs from the richer suburbs to tackle the endemic poverty of the city. The same needs to happen on a much wider scale.

Finally, Northern NGOs will have to think less in terms of doing everything and more in terms of thinking everything. Knowledge, expertise, access to expertise and lobbying is where Northern NGOs can bring most value, not moving bulk goods or supplying semi-skilled labour in emergencies. Northern NGOs as knowledge agencies should be the goal.

The need for independent, people-to-people organizations will not go away. It is a valuable part of the democratic process, counterbalancing the caution and inertia of government. The agency challenge is to remain relevant to the problems and hopes of those they seek to serve.

Chapter 1 Sources, references and further information

An Independent Review of International Aid. *The Reality of Aid 1996*. London: Earthscan Publications Ltd., 1996.

Anderson, Mary. *Do No Harm. Supporting Local Capacities for Peace through Aid*. Cambridge, Massachusetts: The Collaborative for Development Action, 1996.

Bennett, Jon. *Meeting Needs*. London: Earthscan Publications Ltd., 1995.

Edwards, Michael and Hulme, David (eds). *Too Close for Comfort? NGOs, States and Donors*. London: Macmillan/St. Martin's Press, 1996.

Edwards, Michael and Hulme, David (eds). *Beyond the Magic Bullet: NGO Performance and Accountability in the Post-Cold War Years*. West Hartford, Connecticut: Kumarian Press, London: Earthscan, 1996.

International Federation of Red Cross and Red Crescent Societies. *World Disasters Report 1996*. Oxford: Oxford University Press, 1996.

Omaar, Rakiya and de Waal, Alex. African Rights. Various publications. African Rights, 11 Marshalsea Road, London SE1 1EP, UK, tel. (44) (171) 717 1224; fax: (44) (171) 717 1240.

Overseas Development Institute. Various publications. Overseas Development Institute, London.

Richards, Paul. *Small Wars and Smart Relief; Radio and Local Conciliation in Sierra Leone*. In the Creative Radio for Development Conference Report, Health Unlimited, London, 1996.

Richards, Paul. *Fighting for the Rainforest; War, Youth & Resources in Sierra Leone*. London: James Currey, 1996.

Ritchie, Cyril. *The Changing Role and Impact of the UN System: Working with Non-Governmental Organizations*. Paper printed for the Academic Council on the UN System, 1996.

Smillie, Ian and Helmich, Henny. *Non-Governmental Organizations and Governments: Stakeholders for Development*. Paris: OECD, 1993.

Steering Committee of the Joint Evaluation of Emergency Assistance to Rwanda. *The International Response to Conflict and Genocide: Lessons from the Rwanda Experience*. 1996.

Stockton, Nicholas. *Value for Money or Money for Values*. Paper on Humanitarian Standards. Oxford: Oxfam, 1996.

United Nations Department of Humanitarian Affairs. *Aid Under Fire: relief and development in an unstable world*. Geneva: UN-DHA, 1995.

Vincent, Fernand. *Forum Special Edition*. Geneva: IRED, 1994.

Vincent, Fernand. *Alternative Financing of Third World Development Organisations and NGOs*. Colombo, Geneva, Niamey: IRED (Development Innovations and Networks), 1995.

Weiss, Thomas G. and Gordenker, Leon (eds). *NGOs, The UN, & Global Governance*. Boulder, London: Lynne Rienner Publishers, 1996.

Web sites

ICRC: http://www.icrc.org

International Federation: http://www.ifrc.org

ReliefWeb: http://www.reliefweb.int

Relief and Rehabilitation Network: http://www.oneworld.org/odi/rrn/index.html

WorldAid96 (ICVA): http://www.worldaid.org

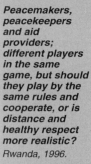

Peacemakers, peacekeepers and aid providers; different players in the same game, but should they play by the same rules and cooperate, or is distance and healthy respect more realistic? Rwanda, 1996.

Chapter

2 Can military intervention and humanitarian action coexist?

Mention of humanitarian action in United Nations (UN) Security Council resolutions has increased 50-fold since 1990, giving the impression that it now features prominently on the global agenda and that solutions to suffering are being urgently sought by a concerned international community. This decade's humanitarian boom reflects the efforts of emergency relief agencies to meet greater demands, but humanitarian action has become a risky business. Relief convoys are blocked or diverted. Aid workers are killed or taken hostage. This is partly due to the disorder allowed by the paralysis of host-nation governance and the collapse of state institutions – an apparent characteristic of the post-Cold War world, but it is also a product of more agencies, from increasingly-diverse backgrounds, seeking an operational role in the chaos and disorder of today's conflicts and wars.

In these situations, the military may appear useful partners to humanitarian action. Many look enviously at the resource base, discipline and "can do" mentality of professional armies. But the reality of Somalia and Bosnia suggest that the linkages between military action and humanitarian action are fraught with difficulties. The association of military and humanitarian organizations in complex emergencies has been beset with problems. Discussion of what support the military can offer to humanitarian operations, and principles to guide it, has failed to keep up with the rapidly-changing environment within which the military and the humanitarian actors work. The challenge of military support to humanitarian operations, however, is less practical than conceptual; the failure of supporting military action has often reflected confusion over purpose, roles and limitations. Valuable practical lessons about military-humanitarian cooperation only become clear when misunderstandings on these issues are disentangled.

The increasing free-for-all of humanitarian agencies and operations in recent years, coupled with the phenomenal growth in outside agencies seeking to operate in other people's countries, have concealed the origins of humanitarian action and obscured its nature. The confusion has had serious practical consequences, particularly when the military have become involved. "Humanitarian" has been loosely used to label motive, problem, task and outcome. Any action or intervention claiming vaguely benevolent motives has sought humanitarian status.

Indiscriminate use of the term may be useful for governments to sanctify their own deeds but for policy-makers, planners, practitioners and academics, it has had a negative impact. At worst it can:

- discredit and limit humanitarian action;

- prevent systematic analysis;

- undermine International Humanitarian Law;

- misuse military resources;

- divert attention away from much-needed political action; and

- foment international discord.

The main losers in this conceptual disarray are, of course, the potential beneficiaries of humanitarian actions: civilians amid conflict, the vulnerable and the poor, women and children, the sick and the old.

This chapter examines the nature of the relationship between military intervention and humanitarian action, focusing on complex emergencies and seeking to establish just what kind of relationship is possible and acceptable between these two very different actors.

Humanitarianism proper

What is "humanitarian action"? It has been shaped by the tradition which gave rise to the International Red Cross and Red Crescent Movement, then to International Humanitarian Law and, in parallel, to most of the major non-governmental humanitarian agencies. Humanitarian action by this tradition requires a neutral and impartial and, by extension, non-coercive way of working. States have codified this in the Geneva Conventions, the International Red Cross and Red Crescent Movement in its seven Fundamental Principles (see inside back cover) and many non-governmental organizations (NGOs) in their mission statements and charters. Action is humanitarian if controlled by the principles of neutrality, impartiality and independence. Relief activity, however well-intentioned or effective, if not conducted in this way is not humanitarian. This understanding of humanitarian action is also incorporated into the *Code of Conduct for International Red Cross and Red Crescent Movement and NGOs in Disaster Relief* (see Chapter 11). If history had been different over the past one hundred years, perhaps another definition of humanitarian action would be with us today. The reality, however, is that humanitarian action, by sticking clearly to the principles of neutrality, impartiality and independence, has allowed life-saving assistance to be delivered in countless conflicts to millions of suffering people. With such a track record, we need to be very sure of our ground before rejecting this tried and tested approach.

Under this traditional paradigm, humanitarian action is apolitical and, by definition, excludes the use of military force. The term "humanitarian intervention", when applied to military action, is thus self-contradictory. The military cannot qualify as humanitarian; every soldier, whether backed by a nation, regional organization or the UN, comes connected to political governance. Military intervention, to achieve political ends, may often be a welcome part of the international community's response to conflict, but it detracts from the clarity of purpose of both the military and the humanitarian to try and fuse the two together in one phrase. Thus, most emergency-relief organizations avoid association with the military because they believe this will prejudice their own neutral and independent status.

This worry over separating military and humanitarian action has not always been so acute. The Geneva Conventions allow for army medical corps to use the Red Cross or Red Crescent symbol and for such corps to treat their own wounded, those of the opposing army and wounded civilians whom they may come across.

But today, it is not being pedantic, naive or timid to seek clearer delineation of humanitarian and military action. It is pragmatism; a concern to keep to a practice with agreed rules that stands a chance of being respected. An immediate benefit of "pure" humanitarian action is access to human suffering even at the most acute stages of a conflict. If relief agencies are perceived as apolitical and non-threatening by the parties to the conflict, they are more likely to gain admission to regions and peoples that would be closed to politically-sponsored organizations.

> *...humanitarian action is apolitical and, by definition, excludes the use of military force.*

The International Red Cross and Red Crescent Movement has gained admission to scenes of acute human suffering in Liberia, DPR Korea, Burundi, Rwanda and the Caucasus at times when access was denied to all agencies with political affiliations, including armed forces. Humanitarian agency experience has repeatedly demonstrated that neutrality, impartiality and independence also normally grant aid workers better protection than armed escorts. The sound practical benefits of the humanitarian approach serve the interests of both aid workers and their beneficiaries.

Cause and effect

Is forceful action to relieve human suffering ruled out? No, indeed, even International Humanitarian Law in Protocol I of the Geneva Conventions allows for such an incidence. The potential of forceful action is revealed when it is properly separated from purely humanitarian action. While humanitarian action can only address the effects of conflicts, politico-military action – provided it does not pose as humanitarian – has the potential in its coercive power to address what humanitarians leave untouched, the causes of human suffering.

Military operations can stop the killing, control violence, secure corridors and police security frameworks. Humanitarian activity can do none of these things, but with limited resources it can reach and mitigate human suffering in inaccessible places and during conflict. Managing these parallel processes of political and humanitarian action is the conceptual challenge of military support to humanitarian operations.

In theory, there are three distinct arenas: political, military and humanitarian. Although military interventions can be seen as extensions of the foreign policy of a state or group of states, in practical terms on the ground they are sufficiently different from other political tools to warrant a separate category. Military and humanitarian agencies each have unique roles in their respective arenas. They are not interchangeable. If the dividing line between these arenas is blurred or crossed then, in fragile security environments, each will compromise the other and neither will be able to act effectively.

The pressures and opportunities in recent complex emergencies have encouraged new concepts of humanitarian action that do blur the lines. Such incoherent and flawed concepts have detached humanitarian action from its traditional origins. They have inhibited clear thinking and brought political, military and humanitarian action together into something of dubious legitimacy and limited effectiveness.

The trend has been for humanitarian agencies to become politicized and for military elements to enter the humanitarian arena. The latter has been facilitated by the tendency of governments to exploit humanitarian operations as substitutes for policy and political action. This has been a mutually-reinforcing error; humanitarians feel obliged to act when the politico-military establishment fails to do so, and the military then become involved in humanitarian activity to protect the compromised, and thus vulnerable, aid workers. Neither group has been served by the exchange. These new concepts of humanitarian action have been practised and found inadequate in several large emergencies for long periods, most notably in Bosnia and Somalia. In the past in less tested environments, such as UN peacekeeping interventions in Cyprus and the Lebanon, cooperative action has been possible between peacekeepers and humanitarian agencies. But in today's more complex environment, it is questionable if such relationships could be repeated.

Box 2.1 **Cardinal ethics and a little jurisprudence**

Two cardinal ethics dominate the legal aspects of war:

- *jus ad bellum* judges the reasons for fighting a war, whether it is just or unjust.
- *jus in bello* judges the manner in which a war is fought.

For two centuries, the focus of law dealing with war has ignored *jus ad bellum* and focused exclusively on *jus in bello*. This was for purely pragmatic reasons, as the international community could rarely be expected to agree on a politically-charged *jus ad bellum*. Legal protection that depended on a consensus about the causes of a conflict would rarely benefit its victims. Essentially apolitical, *jus in bello* was convenient as the basis for International Humanitarian Law, if only as the simplest form of international consensus.

Humanitarianism, according to law, describes a way of doing things and is impartial, neutral and independent in its application. It aims to protect people against the worst consequences of war, seeking to mitigate its effects by limiting the choice of military methods, and by obliging belligerents to spare those who do not – or no longer – take part in hostile actions.

Pragmatism depoliticizes humanitarianism; the concern is to codify law that stands some chance of being observed. The ICRC, for example, expects to be able to carry out its humanitarian operations without affecting any party's legal or political position. A pure humanitarianism, equally and impartially on offer to all in need, demands that *jus in bello* be separated from *jus ad bellum*. The gains of such a separation are indisputable, and account for all the occasions when humanitarian organizations have been able to alleviate suffering amid conflict.

Jus ad bellum and *jus in bello* also relate to the separate cause-and-effect components of emergency. While politico-military action – usually initiated in the context of *jus ad bellum* – can address both cause and effect of human suffering in conflict, humanitarian action – reflecting *jus in bello* concerns – can only address effects. Although political action may therefore relieve human suffering, humanitarian action cannot solve the political causes of that suffering. UNPROFOR's operations in former Yugoslavia exposed the limits of humanitarian action in the absence of effective political will to address the conflict.

In jurisprudence terms, politicizing humanitarianism seeks to reunite the legal principles of *jus ad bellum* and *jus in bello*. Supposedly humanitarian organizations that stridently denounce governments are typical of attempts to do this. Such advocacy rarely acknowledges its disadvantages.

Abandoning humanitarian principles means abandoning the trust of a significant and influential proportion of those in any conflict, and thus also abandoning the unique access to beneficiaries that neutral, impartial and independent humanitarian agencies can achieve, even at the height of a crisis.

Emergency-relief organizations that observe humanitarian principles must defend and sustain them, and the trust they build, or risk being undermined by political activists claiming to share – but fundamentally misusing and misunderstanding – the same label.

*Figure 2.1
Two separate
arenas.*

Politico-military	**Humanitarian**
Diplomatic, military and economic means,	Humanitarian agencies,
undertaking coersive, partisan action,	undertaking consensual, neutral impartial and independent action,
seek to find durable solutions,	seek to mitigate,
to the causes of conflict.	the effects of conflict.

To understand better why humanitarian and military action must be kept separate, and to help seek out ways in which such separate actions can complement, rather than detract from, each other, it is helpful to distinguish four different types of operating environments in which humanitarian and military action can take place: benign, fragile, volatile and conflictual. In essence these involve escalating levels of violence and security risk.

Benign environments

Benign environments can be natural or artificial. The international response to the Kurdish crisis of April 1991 was an example of the latter, the response to the Bangladesh floods in the same year an example of the former. Nearly two million Kurds fled Iraq. Most entered Iran and were helped by the Iranian Red Crescent and other agencies. Those seeking to enter Turkey were blocked at the mountainous border where their worsening condition prompted international concern. Little action was taken by Western nations to insist that Turkey – a party to the 1951 Refugee Convention and its 1967 Protocol, and both a member of NATO and a would-be member of the European Union – allow refugees to enter its territory, where their needs could be met easily and cheaply.

Instead, UN Security Council resolution 688 encouraged US, British, French and Dutch military forces to create safe havens within the sovereign territory of northern Iraq. A large emergency relief operation was launched into those havens, in which military airlift played a key role. The cooperative military-humanitarian venture succeeded because the Western powers were in a unique position to impose terms on Iraq following the Gulf War. This politico-military operation has since been hailed as a precedent of "humanitarian intervention". Only through its application to Bosnia and Somalia has this misinterpreted concept declined in credibility.

In most benign environments, neutrality is not an issue. Unless the military are perceived by their own nation as tools of government oppression, addressing the consequences of a natural disaster will normally make the incompatibilities of military and humanitarian actions largely irrelevant or allow them to stay hidden. American armed forces regularly assist in disaster relief following hurricanes that hit mainland USA. Indian armed forces play a significant role in responding to earthquakes and floods in their country. Some would even argue that in benign environments without arms the military are merely civil servants in uniform.

The Kobe earthquake on 17 January 1995 killed 5,500 people and injured many more. Within hours, medical teams from the Japanese Self-Defence Force were

Figure 2.2
Fragile security
environments

VOLATILE	CONFLICTUAL
Environment unstable but controllable	Widespread and unrestrained conflict
Military activity: **peacekeeping**	Military activity: **peace enforcement**
Separation of military and humanitarian not essential but desirable	Separation of military and humanitarian essential

working in tandem with the Japanese Red Cross Society. Respective military-humanitarian roles in such situations will be determined principally by the location and ownership of human and physical resources. Coordination will be the key to effective performance. Conceptually, military-humanitarian cooperation in benign environments will usually pose no fundamental dilemmas.

Fragile environments

In fragile security environments, where the threat of violence is ever present, an overt military-humanitarian partnership would compromise the status of each and therefore prevent the most effective action in each arena. Neither the military nor the humanitarians would be able to operate freely if openly linked together, and covert links would merely delay and deepen the eventual negative impact. Effective action is likely to depend upon the separation of the politico-military and humanitarian arenas. A twin-track approach will enable military and humanitarian agencies to use their unique capacities to their full potential.

Empowering parallel, distinct but complementary military and humanitarian actions are the key. The incompatibility of military and humanitarian characteristics is not the difficulty, it is the conceptual failure to recognize and allow for such incompatibility. Lack of vigorous analysis by governments has been behind more than one international security catastrophe. To support humanitarian operations in fragile security environments, the military must recognize the need to preserve the delicate humanitarian status of relief organizations. This can only be achieved by distancing itself from them. This conceptual tenet goes well beyond calls to improve humanitarian operations through better military-civil coordination, which misses the point and may exacerbate the problem.

Being distinct may be easier than being complementary at a distance. The military can support aid workers indirectly by their actions but the wider effects of politico-military activities at the strategic or operational level can hinder aid.

Equally, humanitarian action can create side-effects that are unhelpful to the politico-military conflict resolution processes. Roads improved by the UN in Bosnia to ease access for food and medical convoys also made it easier for factions to move troops and guns. In many fragile environments, aid brought in by humanitarian agencies can be one of the most valuable resources in the area. It can become fought over and fought for. It can inadvertently fuel violence. Diverted food and fuel can power military vehicles or feed fighters.

When the military and humanitarian work in parallel in conflictual environments, every effort needs to be made to relate the effects of their separate actions to a common aim. Once free of conceptual confusion, there is much that the military can do to support humanitarian operations in ways that are complementary but distinct, and without prejudicing the status of emergency-relief agencies.

Volatile environments

Volatile environments – unstable situations which stop short of widespread conflict – will usually require military forces to undertake a wide variety of peacekeeping tasks. There is risk in open military-humanitarian association; whatever the immediate consequences, it will make humanitarian agencies dependent on continuing stability and the success of conflict resolution.

Armed escorts for aid workers are not likely to prove effective. Use of locally-enlisted armed guards to protect relief deliveries within Mogadishu met a short-term need but their long-term effect was severely counter-productive, turning as they did into a classical protection racket. Perceived impartiality, neutrality and independence will guard humanitarian agencies more effectively than close protection, especially if the military are committed separately to establishing a wider security framework.

The common aim of military – usually peacekeeping – activities in volatile conditions is to stabilize and improve the prevailing environment. Forcible pacification, even if possible within the usually limited resources of peacekeeping missions, cannot be a practical long-term solution. Effective peacekeeping requires the cooperation and

---

Box 2.2 The risks of military-humanitarian linkage

Military peacekeeping in volatile situations serves indirectly, at least, to improve the operational environment. This can be a vital contribution to humanitarian endeavour. Promoting and preserving consent is the guiding principle of military peacekeeping. If consent is maintained and the operational environment remains stable, humanitarian action stands a chance of remaining largely unopposed, even if identified with the military contingents of a peacekeeping operation.

However, association between military and humanitarian operations carries risks. Should the peacekeeping contingent lose local consent and support and be confronted by opposition, humanitarian operations that have become associated with the military may also face resistance. By linking themselves to the military, the status of emergency relief workers becomes dependent upon the success of the peacekeeping mission. If it fails, the humanitarian operation is likely to fail with it.

Humanitarian operations in Bosnia under the protection of UNPROFOR demonstrated this. UNPROFOR worked hard to develop and maintain a consensual environment as the most basic necessity of its mission to protect the delivery of humanitarian aid. Much was achieved. The UN claimed that humanitarian assistance saved hundreds of thousands of lives.

However, when UNPROFOR initiated offensive action by NATO air strikes, peacekeepers and humanitarian workers alike were taken hostage or killed, and humanitarian operations were severely cut back, demonstrating the vulnerability of military-humanitarian linkage. In a similar way, local hostility aroused by American military operations in Mogadishu in 1994 endangered the activities of humanitarian agencies because of their direct association with UNOSOM II.

These experiences subsequently caused IFOR to fight shy of close involvement in humanitarian activities in Bosnia despite the relatively-stable environment. The potential for consent to be lost in peacekeeping operations suggests that humanitarian agencies should preserve their independence from politico-military conflict resolution processes.

It is not just military linkages that humanitarians need to treat with caution. Rwanda's Special Representative of the UN Secretary-General in early 1996 regarded judicial redress as the key to national reconciliation that would prevent renewed outbreaks of genocide. But collating and passing on information for court cases is not the job of humanitarian agencies. Like military forces, human rights agencies and tribunals have had to understand that humanitarian organizations must maintain an actual and perceived distance.

---

consent of the majority of the local population and the leadership of the principal authorities, be they belligerent factions or government agencies. Traditionally, intervening military forces have seen consent as a matter between them and the military power controlling the area within which they were to operate. But today, consent, whether for military or humanitarian operations, is taking on a wider meaning, with both sets of actors valuing more highly the opinions and fears of the local population as well as the controlling powers.

If humanitarian operations, which are also based upon the principle of consent, are to be preserved in volatile conditions, military peacekeeping contingents must avoid damaging the consensual environment. A negative example of this was the expanded UN Operation in Somalia (UNOSOM II) in Mogadishu. The hunt for faction leader General Aideed in 1993 was conducted with little apparent regard for its deeply-unfavourable repercussions on the operational environment and the activities of humanitarian agencies.

Peacekeeping tasks. Peacekeeping is the common task of the military in volatile environments. Even a rudimentary security framework is hugely valuable to humanitarian activities. Without it, the long-term success of any emergency relief is problematical. Military peacekeeping tasks typically include:

- conflict prevention;
- demobilization operations;
- military assistance; and
- guarantee or denial of movement.

Conflict prevention, by preserving a relatively-secure environment, provides a situation that will facilitate humanitarian operations. It can involve up to four complementary activities:

- early warning;
- surveillance and intelligence;
- stabilizing measures; and
- preventive deployment.

Demobilization operations will enhance the security of relief activities through the controlled withdrawal, disarming and rehabilitation of belligerents with prior agreement as part of implementing agreed settlements. Demobilization and humanitarian activity have successfully progressed in parallel in both Mozambique and Angola, the former involving 90,000 government and Renamo forces during the period from 1992 to 1994. Both delays and premature action in demobilization may threaten stability and endanger relief-agency operations. At several Mozambique garrisons, delayed demobilization led to disorder that threatened aid operations.

Military assistance covers all assistance given by a military under its peacekeeping mandate to a civil authority, from supervising a transfer of power to reforming security forces and developing or supporting civil infrastructure facilities. Maintenance of law and order is fundamental to such activity. These tasks will facilitate humanitarian operations, although their nature risks overt linkage with emergency relief agencies.

Guarantee or denial of movement can prove crucial to the conduct of relief work, particularly delivery of food or evacuation of the wounded, such as operations to allow ships, aircraft or trucks to reach a besieged city. Denial of movement may focus on establishing no-fly zones over a region. Guarantee and denial measures may overlap with conflict-prevention measures, including preventive deployment.

Box 2.3 Use and preservation of consent in peacekeeping

The presence or absence of consent represents the crucial difference between volatile environments (in which the military undertake peacekeeping) and conflictual environments (in which the military undertake coercive peace enforcement). If a peacekeeping force comes into a volatile environment with the consent of the different parties to the conflict, its operational conduct should seek to preserve such consent.

There are various ways in which consent might be inadvertently breached (top figure).

Successful peacekeeping – and self-protection by peacekeepers – depends on the exercise of principles to protect the force's neutral referee status. Critical peacekeeping principles emerge by considering counters to inadvertent breaches of consent (bottom figure).

Taking sides		
Too much force		
Loss of legitimacy	Absence	
Loss of credibility	of consent	
Gross disrespect		
Misunderstanding		

Taking sides		Impartiality
Too much force		Minimum force
Loss of legitimacy	Consent	Legitimacy
Loss of credibility		Credibility
Gross disrespect		Mutual respect
Misunderstanding		Transparency

Conflictual environments

If conflict is, or becomes, widespread, peacekeeping will have to make way for peace enforcement, which is not dependent upon consent but seeks to coerce by force. Conflictual environments do not rule out humanitarian operations. In these situations relief activities can achieve unique success, gaining access to situations of human suffering denied to even heavily-armed military contingents.

The International Committee of the Red Cross (ICRC) has demonstrated this in much of Afghanistan, the UN Children's Fund (UNICEF) in south Sudan. Access usually depends on the perceived "purity" of the agency's humanitarian status. If relief organizations are believed to be truly humanitarian, by being – and being perceived to be – neutral, impartial and independent, they stand a reasonable chance of acceptance by belligerents.

Linking military and humanitarian agencies in such situations – or permitting the perception of links to grow – will politicize aid workers and deny them the critical status essential for access. If humanitarian agencies are perceived as taking sides they may face attack, and if the military are committed to support relief activities, they will also be made more vulnerable. The UN Protection Force (UNPROFOR) in former Yugoslavia, for example, allowed itself to be scattered as it tried to cover dozens of small humanitarian operations at once and protect the "safe areas". Its commitment to protecting the delivery of relief supplies demanded weak, unbalanced deployments that prevented concentration of force and more proactive military tasks.

While the focus of humanitarian operations in conflictual environments is human suffering, the cause of that suffering will also demand action. The proximate cause of suffering is likely to involve armed disputes and addressing them will usually require politico-military tools and coercive, partisan action, the sort of action humanitarian agencies can never take, but the sort of action which peacemaking forces can indulge in. Such action is very much in the interests of the people humanitarian agencies seek to help. In conflictual environments, parallel but distinct activity by military and humanitarians will represent a comprehensive approach: the military, if properly resourced and mandated, can address the proximate cause, while humanitarian agencies can mitigate the effects.

The military could use force to gain access, but this requires enormous resources, is likely to be short-lived, risks significant casualties and could well have a negative long-term impact on beneficiaries. By early 1993 in Somalia, the neutrality of humanitarian organizations was undermined by their enforced close association with, and subordination to, the American-led Unified Task Force (UNITAF). By July 1993, the expanded UNOSOM II was at war with one party to the conflict, with disastrous consequences for all. One former Commander of UN Bosnia-Herzegovina Command believed only an invading force given complete authority to take over an entire country could deliver aid without consent.

Conflictual environments are the most complex situations in which to consider the potential for military support for humanitarian operations. Even so, there are four possible ways in which the military might support humanitarian operations in such situations:

- military "circuit-breakers";
- forcible emergency relief;
- creating humanitarian space; and
- subcontracting military assets.

Circuit breaking. Given the right conditions and political backing, military forces may sometimes be able to transform security environments by short actions that break cycles of violence, exerting sudden radical influences on the prevailing balances of power. Significant changes in the security situation can offer a return to peacekeeping-type environments. It is unlikely that such actions would succeed without the backing of the UN and one or more major powers. Circuit-breaking offers a means to create an operational environment that favours humanitarian action.

In Cambodia in 1993, for example, when a Dutch peacekeeping platoon came under attack, they called in support in the form of an entire marine battalion; following a brief but powerful exchange of fire, the area of the attack became secure and safe. Whilst this example is tactical rather than strategic, it hints at what might be possible given sufficient political will.

Forcible relief. The notion of forcibly delivering emergency relief to suffering peoples has popular appeal. However, such action will involve practical difficulties requiring resources and political will rarely available, and unlikely to be sustained. Such action has no relationship to humanitarian action. This concern is not pedantic but practical. Disguising force, however worthy its motive, as humanitarian undermines genuine humanitarian actions. It will also foster discord and suspicion within the international community, which might sometimes suspect ulterior political motives. Forcible relief in the absence of other aid efforts could be both effective and welcome. However, if undertaken in association with humanitarian agencies, it puts their long-term work at risk.

The concept of humanitarian space represents a classic distancing technique employed to carefully separate military and humanitarian action, and offers considerable potential for effective military-humanitarian synergy when working in conflicts. The role of armed forces in this situation is to provide physical security and freedom of movement for all, such as keeping airports and roads open, and carrying out mine clearance. Such a security framework could enclose a protected zone within which impartial relief agencies would operate in a neutral and independent fashion. It could also involve relief corridors to centres of population.

The use of lightly-armed UN guards in Iraq in 1991-92, and the use of the flight-exclusion zone in northern Iraq from 1991 onwards, both created space within which humanitarian action was able to more safely take place. The Head of Civil Affairs in Bosnia described the 1994 establishment of heavy-weapons exclusion zones around Sarajevo and Gorazde as a military operation creating conditions under which human suffering could be alleviated, emphasizing that such a use of military force was entirely different from humanitarian operations. Ideally, isolating humanitarian space from the security forces creating it should preserve the perceived impartiality, neutrality and independence of humanitarian agencies working within it, thus safeguarding their access and protection. Bosnia, tragically, has also provided examples of the failure to properly secure humanitarian space. Many of the so-called safe zones were anything but safe.

Creating humanitarian space goes far beyond identifying a geographical area and securing a perimeter. It need not even be created by military forces, but through more intangible means, such as diplomacy and military threat. The discreet presence of warships may be enough to create the "space" needed by relief workers inland. Measures that uphold law and order and undermine "cultures of impunity" also create humanitarian space.

The need for distinctive action does not prevent governments subcontracting to humanitarian organizations a range of military equipment and personnel in such sectors as transportation, including airlifts and airport handling services; logistics, warehousing and fuel provision; engineering and construction, including water supply, sanitation and ordnance disposal; medical, such as field hospitals; and communications.

Although military resources are usually far more expensive than their civilian equivalents, this support can greatly enhance humanitarian capacity. However, governments do not usually want to lose control of their military assets in this way, especially if significant risk is involved, which limits state contributions of military assets. Legitimacy demands that such resources should be transferred to the control of relief agencies while the importance of perceptions mean that subcontracted military resources must be "distanced" from the humanitarian agency receiving them. One suggestion is for military equipment to be lent or leased to a humanitarian agency along with retired military staff who are capable of operating the equipment but do not carry the political strings of serving officers.

For example, a US military Starlifter aircraft with an air-force pilot can only be associated with the Red Cross through its virtual transformation into a civilian plane. It will need to be painted white with Red Cross markings, call signs and radar code will need to be altered, no weapons will be allowed on board and there will be limited use of military uniforms during duty time. If such resources are not clearly under civilian command and control, however, they carry the risk of creeping politicization.

Conclusion

Military intervention and humanitarian action can coexist. Military intervention can increase the likelihood of success of humanitarian action.

In benign security environments where the issues of neutrality and consent are not questioned and are, for the most part, an irrelevance, military operations can have a very close association with humanitarian action. The relationship jeopardizes neither the immediate missions of either party nor their future perceived ideological purity.

As we move up the escalating scale of violence from fragile to volatile to conflictual security environments, the relationship becomes more awkward and greater and greater distance is needed between the two sets of actors in order to allow each to perform their tasks effectively. Thus military and humanitarian agencies are best employed in applying their unique capacities to their separate respective arenas in a complementary but distinct fashion. Both proximate cause and effect of human suffering can then be addressed in parallel.

To achieve such a synergy the following conditions would need to be met:

First, relief agencies operating in violent environments must practise and be seen to practise neutrality and impartiality and, as a consequence, to operate with the consent of the population.

Second, military interventionists, particularly force commanders, must have a clear understanding of the purpose of humanitarian action and the rules by which agencies play. Thus understanding must also be reflected in the mandate given to the intervening force.

Thirdly, since both sets of actions share in part the common goal of alleviating suffering and since both sets of actors can either complement or detract from each other's mission through their actions, there has to be a clear commitment to regular and open communication, both on the ground and at the political level from which military and humanitarian actions are directed. On the ground, in particular, an atmosphere of trust has to be built up between the humanitarian and military actors.

A body of experience in the successful military and humanitarian management of twin-track operations is only slowly being accumulated. But it is clear from many past mistakes that in such situations humanitarian action must carefully sustain its impartiality, neutrality and independence, while the best use of the military is in exploiting its politically-backed coercive power to control violence and protect life.

Chapter 2 *Sources, references and further information*

Belgrad, Eric and Nachmias, Nitza. *The Politics of International Humanitarian Operations.* Praeger, 1996.

Best, Geoffrey. *War & Law Since 1945.* London: Clarendon Press, 1994.

HMSO. *Wider Peacekeeping.* London, 1995.

Humanitarian Peacekeeping. *Global Society.* 1996.

International Federation of Red Cross and Red Crescent Societies. *World Disasters Report 1995.* Dordrecht, The Netherlands: Martinus Nijhoff Publishers, 1995.

International Peacekeeping (quarterly journal), Frank Cass, London.

Macrae, Joanna and Zwi, Anthony. *War & Hunger.* London and New Jersey: Zed Books, 1994.

Ramsbotham, Oliver and Woodhouse, Tom. *Humanitarian Intervention in Contemporary Conflict.* Polity Press, 1996.

Whitman, Jim and Pocock, David, eds. *After Rwanda: The Coordination of United Nations Humanitarian Assistance.* London: Macmillan, 1996.

Whitman, Jim. *A Cautionary Note on Humanitarian Intervention.* GeoJournal 34.2, pp. 167-175, October 1994.

Web sites

Humanitarianism and War Project: http://www.brown.edu:80/departments/Watson_Institute/H_W/

International Federation: http://www.ifrc.org

Journal of Humanitarian Assistance: http://www.jha.sps.cam.ac.uk

Section Two Methodologies

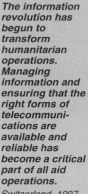
Chapter

3 From information management to the `knowledge agency'

Humanitarian organizations are in the information business. As clearing houses of data, they collect, analyse, store and communicate facts and figures. Information in disaster relief is a primary tool and an essential resource, it translates into supplies, logistics, and agency cooperation. Speed, accuracy, and completeness of data can help save lives.

Improving information management could not only assist agencies to do their job faster, cheaper and better, it offers the opportunity, perhaps the necessity, to transform what existing humanitarian agencies do, how they do it and many of the present relationships between the variety of stakeholders in disasters, from governments to millions of beneficiaries.

At a time when Northern donor disengagement is cutting aid budgets despite growing vulnerability, forcing ever greater efficiencies upon agencies, information technology – which can help cost-effective use of resources – is one of the few disaster resources falling in price.

Meanwhile, the plunging cost and soaring scope of computer hardware, information software and telecommunications increasingly offers disaster professionals in the developing world – and, ultimately, disaster victims themselves – the chance of taking far greater control of all aspects of crisis management, from relief operations to preparedness programmes.

Box 3.1 Improving services with integrated information in Sierra Leone

In the hope of peace after years of internal conflict, the government of Sierra Leone established the Ministry of National Resettlement, Rehabilitation and Reconstruction (MNRRR) to coordinate a range of programmes, including the resettlement of approximately 1.2 million internally-displaced people (IDP) and refugees.

In 1996 the World Bank funded Response.Net and Volunteers for Technical Assistance (VITA) to develop and deliver an integrated information system needed for the effective coordination of demobilization, resettlement, reintegration and rehabilitation activities. Response.Net and VITA are NGOs using technology and information-management services to help deliver humanitarian services to vulnerable people.

The Integrated Information Centre (IIC) system for MNRRR, using commercial off-the-shelf software, provides communication tools to satisfy the messaging and information-exchange requirements of decision-makers within UN agencies, NGOs, and government organizations in Sierra Leone. It aims to help:

■ coordination and monitoring of programmes by MNRRR;
■ fast, effective information exchange by government, donors, UN agencies, NGOs;
■ accurate reporting to donors;
■ identify information needs of government and non-government agencies; and
■ build up information capacity to feed into future development and economic growth.

The IIC is located at the MNRRR in Freetown and is staffed by two specialists with expertise in both systems operations and information management. It is designed for ready access in Freetown initially, and to be accessible throughout Sierra Leone in the future.

The key IIC application is a software package that, as well as being Internet compatible, provides direct access via a phone connection using a standard modem device for e-mail, file transfer, database query and on-line browsing of IIC topic areas (such as NGO situation reports). The system has these features:

■ e-mail within Sierra Leone, and internationally when Internet is available;
■ file-transfer capability, to exchange electronic files within Sierra Leone and internationally;
■ topic area, to review and comment on humanitarian issues;
■ capability to connect and interact with an Internet service provider, when available;
■ remote connectivity for database access and processing, including on-line queries; and
■ digital map system, with country datasets able to display working data to view or print.

The IIC offers several services to help operational coordination by individuals and agencies.

One section allows organizations to post their situation reports, including the UNDHA monthly consolidated report. Another section is a "fact book" containing timely information on security, travel, infrastructure, socioeconomics and demographics. Regularly updated by participants, a contact list gives current addresses and relevant information. A discussion area is available to post comments or questions under user-defined subjects, such as IDP/refugees, health, camps, water, food, demobilization, resettlement, reintegration, and rehabilitation.

Allowing on-line queries, databases run by UN agencies, government ministries, and NGOs are "mirrored" on the IIC system, such as the Food Technical Committee database coordinated by the World Food Programme, the UNDHA displaced persons registry, and the Sierra Leone government Statistical Unit of Demographic Information.

The effectiveness of the IIC system in Sierra Leone is not ultimately dependent on the sophistication of its technology, but how it is used by individuals and agencies for coordination, cooperation, and accountability to improve services for vulnerable people.

That requires leadership by those within UN agencies, NGOs and the Sierra Leone government to maintain operational independence yet promote information interdependence, and thus their ability to maximize effective operational activities with each other.

Amid today's often overwhelming information flows, data quality is crucial. Humanitarian agencies will increasingly be judged on the quality of information they produce, and on their ability to synthesize useful data and apply it in real-time relief. However, most agencies are only slowly developing the infrastructure to undertake effective information management.

In this they match the commercial sector, where, as Thomas Davenport argues in *Process Innovation,* few "have treated information management as a domain worthy of serious improvement efforts...vast amounts of information enter and leave organizations without anyone being fully aware of their impact, value, or cost."

> *... information technology – which can help cost-effective use of resources – is one of the few disaster resources falling in price.*

While the reactive, crisis-management focus of relief organizations may help explain the lack of emphasis on information management, they clearly do see good communications, from satellite phones to electronic mail, as essential for linking with their workers in the field, facilitating operations and improving security.

Despite technological improvements, many decisions are still taken in emergencies with very little information beyond that in peoples' heads. There has been little emphasis on identifying what information is essential in relief, which stakeholders need it, how it should be used, or the way new technological tools can facilitate systematic data collection and communication.

Who needs information?

Donor institutions require high-quality data and analysis, including trends and comparative information, and they want it quickly. Aid workers need immediate data about the dispatch of supplies, for example, or information on the best location for a refugee camp. Agency desk officers want fast access to information on personnel with specific expertise, and much more.

Each disaster stakeholder has very different information needs and priorities, as well as having their own influence on its management. Low understanding of what each stakeholder requires is one reason why systematic organization of disaster-management information has not kept pace with advances in agency use of information and communications technology.

Developments in radio, cellular and satellite communications all facilitate the use of electronic mail, for example, which can help answer specific questions or resolve a particular problem. The International Committee of the Red Cross (ICRC) is using the Internet to improve the tracing of missing people from Bosnia. Its directory of 11,000 people is simply an alphabetical list, but the Web site permits searching by date, location and other factors. Both those with information and individuals listed as missing can send e-mail messages through the site.

Yet, in general, disaster-management information on logistical issues, resources, and specific relief action is only slowly becoming available in dynamic systems. For example, as the Rwanda crisis unfolded in 1994, humanitarian organizations were desperately seeking French-speaking aid specialists to set up operations. If up-to-date information sources focused on these needs had been available globally, this search would have been much simpler.

Much predictive data for natural disasters – from meteorological satellites or the mapping of seismic activity – is available, as is information on man-made and technological hazards, and both are being used by emergency-management professionals. However, the humanitarian community needs an enormously-diverse range of information, such as:

■ availability and movement of relief supplies;

■ population displacement;

■ capacity of local infrastructure, such as ports, hospitals and airfields;

■ disease surveillance, including baseline epidemiological information (see Chapter 4);

■ specialist, relief expertise available;

■ data on other agencies involved;

■ national and local facts and figures on the economic, social and political situation; and

■ mapping and other geographical information.

Information needs vary widely during the different phases in disaster management. For example, information required at the disaster-preparedness phase includes:

■ country-specific information, such as demographic and geographic data;

■ emergency plans and national development plans;

■ reports from other agencies;

■ in-country capability, including financial, material and human resources; and

■ documented information from previous disasters, such as recovery programmes.

At the initial stage of disaster response, relief agencies carry out a needs assessment, usually by sending an expert or team to the affected area. Some agencies use assessment manuals as a guide to ensure that key elements are covered. At present, needs assessment is usually the most information-intensive period within the disaster-management cycle.

Cyclone damage

Data gathered through this process is used for drafting funding appeals, and to plan the disaster response, including determining what supplies and expertise are required. Its goal is to quickly obtain an overall idea of the extent and severity of the effects. In urgent situations, agencies may send supplies before the assessment is completed.

The needs-assessment survey can be affected by factors beyond the control of the assessment team. Extensive earthquake or cyclone damage can make an accurate assessment difficult. Movement may be limited and communications poor. Decisions in disasters are frequently based on the information available rather than on the information actually required.

The *Handbook for Delegates* issued by the International Federation of Red Cross and Red Crescent Societies defines four important roles for needs-assessment surveys:

■ assess victims' needs and community capacity – such as local agencies – to meet those needs;

■ systematically analyse this information;

■ recommend any action needed and its form, taking into account local capacities; and

■ prepare an operational plan and budget.

Box 3.2 ReliefWeb: global information system for humanitarian emergencies

UNDHA began building ReliefWeb in 1994 as an international global network for communication and information support.

ReliefWeb aims to be the principal repository of information to assist the humanitarian community in managing emergency response through provision of timely and reliable information on situations, stakeholders and resources.

Its elements include an online information system available through existing networks; links with regional and field-based information centres able to report on complex emergencies; promotion of common standards for reporting and information collection, analysis, exchange and dissemination; and a system of alerts to ensure prompt and effective response.

Its features include: twice-daily publishing of time-critical information; map centre; emergency assets; financial tracking of humanitarian contributions; personalized "what's new" to individual users; and links to useful web sites.

The next phase will see the implementation of linkages with field operations and closer working relations with regional information centres, such as the Integrated Regional Information Network in the Great Lakes, and early-warning and conflict-prevention sources.

ReliefWeb is a cost-effective use of the information highway. It hopes to draw effectively, efficiently and cooperatively on existing resources of agencies and NGOs to:

■ strengthen coordination of humanitarian and disaster relief assistance of the UN system;

■ support inter-agency relief coordination by ensuring access to system-wide information supplied by agencies, inter-governmental organizations and NGOs; and

■ respond to donor requests for greater transparency and accountability on tracking contributions for humanitarian assistance.

Though it cannot itself deliver solutions, with appropriate support and resources it can help users find information whose timeliness and scope support decision-making at all levels.

Some information collected will be irrelevant, while other data may be inaccurate because the sources, such as the government, want to exaggerate or underplay the scale of the disaster. All data needs verification, including that from local organizations. It is hard to produce a needs-assessment survey in which all information is accurate, quickly transmitted and well coordinated. This inevitably hampers the relief operation and inter-agency coordination.

An example of improved information management and new technologies in assessment is the development by the Save The Children Fund-UK, backed by the European Union, of risk-mapping computer software for food and famine. Combining a wide range of data – such as livestock prices, rainfall and harvests, coping strategies, household budgets and the potential impact of conflict – the software helps predict the location and level of food deficits. This should help food-aid targeting, and offer evidence to bring additional international assistance.

It is generally agreed that a systematic method of data collection will help in compiling more accurate reports and in providing information for future use. If data is recorded in a uniform way, it would allow the creation of an information system with needs-assessment information from previous disasters. Comparative data would be easier to identify. Unfortunately, the actual development of common standards for data collection and analysis has been slow.

Systematizing the collection of data has been low on the priority list of humanitarian assistance organizations largely because of the lack of information-management expertise and procedures to handle data. Past meetings of non-governmental organizations (NGOs), United Nations (UN) agencies and others to discuss common standards have generally either gone into far too much detail on how and what to collect – resulting in a lack of commitment by agencies which do not see the added-value for their work – or lacked focus on the specific types of information that are key to relief work.

Conclusions are vital about the latter for progress to be made. Clearly much information in disaster relief is ad hoc by its nature and only part of the data should be routinely collected. But by whom? The collection and standardization of information needs to be part of the job description of specific aid workers as part of their mandate to coordinate relief work.

Situation reports (sitreps) are an important information tool, particularly for the integration phase when recruitment of expertise and channelling of supplies has to be carried out quickly and efficiently. Sitreps are widely disseminated to give stakeholders timely data on the situation, describe how resources are being used, indicate further needs and identify possible future trends. However, few agencies use sitrep data for future planning or post-disaster comparative studies, in part because sitreps are not necessarily structured for reuse of data.

In the production phase of disaster response, operations become routine, requiring basic financial, logistical, administrative and personnel information. Clear procedures and the use of appropriate information tools and systems are vital for these information-intensive tasks.

In terms of information needs, the transition phase is similar to preparedness. As the relief operation is closed down, responsibility for any further work is usually handed over to local agencies, such as national Red Cross or Red Crescent societies. Information needs to relate in particular to training information, both for specialist and management functions, and procedures of various types. The emphasis here is on hand-over: building local organization capacity and infrastructure, in which information plays an important role.

Information workers, too?

Donor governments frequently advocate close coordination and information sharing, yet neither this urging nor new technological tools have brought much improvement. In recent years, donors have funded projects which promised to develop systems and standards to improve global humanitarian information exchange. Unfortunately, these ventures primarily focused on the technological aspects of information dissemination rather than on the actual needs of the relief community in the field or at headquarters.

"Although a real commitment to information sharing is probably the most important single step needed by relief practitioners," the Technical Coordinator of the UN Emergencies Unit for Ethiopia argued in a recent paper, "other issues that need to be addressed include agreements on standards, information modules or structures; in other words how can we share meaningful, relevant information and adjust our information sharing so we meet the real needs of various players."

Emergency organizations are essentially too busy with their main mandate: disaster relief. Most do not have information professionals on their staff to analyse and develop the mechanisms to collect and organize information in a systematic manner. These tasks are often not seen as "core" to the work of the organization. Expectations are often naively high for the use of technology, since these are tools not solutions. Major work is needed to analyse the information needs of each stakeholder and to generate commitment to information management at all levels within humanitarian organizations.

New procedures will need to be developed to ensure focused information collection for these specific requirements. Information management is one part of a significant change in operating culture from amateur relief worker to career professional. Information skills will need to be added to some relief job descriptions, and those workers will need regular training.

One key factor will be enhancing the information-technology infrastructure and capacity to use data within indigenous organizations. This is the aim of an International Federation project launched in 1996 with funding from the Canadian International Development Agency (CIDA) to help National Societies in the disaster-prone countries of Africa and Asia.

The CIDA project focuses on the development of a global network for the International Federation to support disaster preparedness and emergency-response activites, to enhance the essential information-infrastructure capacity and services within National Societies, at the local, national and regional levels. It involves setting up pilot projects in disaster-prone regions, concentrating on both information management and data-communications questions. The goal is to facilitate the exchange of essential operational information in the International Federation, with its partners and other humanitarian organizations through the provision of up-to-date, appropriate telecommunications and information-resource facilities, expertise and technical assistance.

It is essential to ensure that information collection is not cumbersome, a challenge in which information systems and technological tools can help. If information systems are seen as a burden, they will not be used either to collect or communicate information.

What will it cost?

Information companies, such as Reuters, flourish because they invest heavily in expertise and technology. Much of their work is synthesizing data and packaging it for different target groups. Their clients will pay for good quality, reliable and timely information to assist decision-making which produces profitable results. The analysis and synthesis of information – a skill little known among the humanitarian community – is becoming a key factor in successful fundraising, one area where non-profit organizations have a "bottom line".

High-quality information is one of the most valuable, scarce, and thus often the most costly, resources in disasters. It is not just information but knowledge and learning that can help the humanitarian community to do a better job and to improve performance and prioritization. Knowledge implies a sharing of information that has been selected, evaluated and found useful, which in turn implies building an intelligence foundation within disaster-management organizations.

The costs will be for infrastructure, technology, training and expertise, including increased rewards and recognition to attract and retain skilled staff. Without this investment, humanitarian agencies will continue to work in a reactive manner. This will prevent them benefiting from good practice, innovative techniques and lessons learned, and leave them unable to pass on skills and knowledge to partners, especially indigenous organizations.

However, there are some signs that the disaster community has understood the importance of information and will evolve its practices and procedures to give this area the attention it requires. Apart from the International Federation's own focus on information, backed by a major donor government, the World Bank support for the Integrated Information Centre of Sierra Leone (see Box 3.1) and the creation of ReliefWeb (see Box 3.2) by the UN Department of Humanitarian Affairs (UNDHA) are new information initiatives that can demonstrate what is possible.

In the present industry-wide efforts to develop and demonstrate "better" aid, the desire for standards will mean creating the equivalent of more bottom lines than just fundraising income, establishing many more ways to judge success or failure, costs and benefits. Clear indicators will help disaster managers demonstrate how information helps provide more appropriate, timely and cost-effective aid to the intended beneficiaries, and good information management will help develop and implement higher standards.

In time, quality benefits will be quantifiably demonstrable, which will save more lives, cut costs, and make a substantial difference for beneficiaries. This will also have an impact for agencies able to deliver the desired results. In a world of inter-agency rivalry, commercial competition, falling donor budgets and increasingly-sceptical media scrutiny, humanitarian organizations that appreciate the urgency of this win-win proposition will respond by making information management one of their core tasks.

> *Beneficiaries may not wait for agencies to bring in knowledge systems.*

There may well be other costs, if organizations are not prepared for the changes that often follow when information – classically a source of authority and power, even within flat-management NGOs – flows in different ways, through developments such as the internal networks known as intranets, or the concept of data warehousing, which concentrates, integrates and makes available information to increase its value.

While some might see information technology as a way for senior staff to micromanage around the world in the style of telemedicine, present developments are putting far more information into the laptops of field workers. As well as instant information on decisions at headquarters or port deliveries, the data include local facts and figures that were once hard to find. Improvements in communications have been one factor in the trend to shift decision-making forward from headquarters towards the field through agency regional offices.

Discussing information in commercial companies, Jonathan Barling-Twigg, a consultant in enabling technologies at KPMG Management Consulting, warned that intranets and other developments could produce entirely-unplanned changes in their internal power structures.

He added: "They remove traditional and unwelcome controls often used to protect internal information empires and slow change. For certain departments or individuals, however, the changes an intranet bring can be less welcome. An intranet can suddenly remove blocks on the flow of information that traditionally protected administrative dominions from interference and awkward questions...the management of the organization may resent, or feel threatened by, the more proactive, self-reliant activity encouraged by the increase in communication between employees."

Of course, aid agencies investing in open information and communication systems could also encounter more proactive, self-reliant activity in disaster zones encouraged by the increase in communication between their staff, vulnerable people and indigenous disaster professionals.

Beneficiaries may not wait for agencies to bring in knowledge systems. As war spread and telephone lines were cut within the former Yugoslavia, communications activists created a "cyber Bosnia" in the ZaMir ("for peace") Transnational Network (ZTN). ZTN linked cities, groups and individuals within the war zone and outside via relayed links elsewhere in Europe, carrying e-mail, refugee mail, electronic conferencing and

Box 3.3 **Telecommunications and networking in disaster relief**

Today's global telecommunications environment provides a rich matrix of technologies for use in disasters. A major challenge is choosing the appropriate technologies, and balancing factors such as speed, ease of deployment and maintenance, operating cost, technical complexity, redundancy, reliability, and resilience.

These technologies can be considered as four interrelated and overlapping groups:

1. Radio frequency-based transmission systems provide increasingly reliable, flexible and portable local and global communications through a combination of small radio handsets, vehicle-mounted portables, briefcase-size satellite terminals, or high-capacity transportable satellite ground stations. These technologies ensure mobility and continuous functioning when ground-based infrastructure is absent or destroyed by disaster. Developments include:

■ Global Positioning Systems (GPS) units for vehicle tracking, fleet management, location-fixing and personal security. As a low cost add-on to high-frequency radio transceivers used in vehicles, they allow instant identification of a specific vehicle and transmission of its position or an emergency call.

■ Notebook-size 2.5kg INMARSAT Mini M voice/data terminals.

■ Cellular telephone services or an even newer technology – Personal Communications Systems – to provide "infrastructure" where the traditional telephone system is limited.

■ Lower cost and rising availability of satellite-based telemetry, imaging and survey facilities.

■ Availability of more powerful digital data systems using high-frequency radio equipment and capable of being linked to other data networks (e.g., the Internet or local area networks).

2. Traditional ground-based transmission systems for voice telephony, fax or e-mail, which still provide an underpinning for much relief communication, are improving. Increasingly these systems are interconnected to, or supplemented by, radio-based systems. Trends include:

■ Progress towards digitalization (e.g., in transmission and switching) worldwide, especially in developed countries. Rural telecommunications and the "last mile" continue to be problems.

■ Extensive deployment of high-capacity fibre optics links, especially on trunk routes and over long distances (e.g., transoceanic).

3. Data networks make the world increasingly interconnected. Once mostly run by large organizations, particularly multinational corporations, banks, airlines and, in some settings, the military, such facilities are today used by individuals, small groups and firms, and non-governmental and humanitarian agencies. Trends include:

■ Explosive growth of the Internet. With a wide range of information dissemination and communications applications, the reach of the Internet is rapidly becoming global. Even areas once unable to participate (e.g., sub-Saharan Africa) are gaining some access.

■ Local area networks (LANs) enable agencies to work more effectively together. Wireless LANs (using radio frequencies or infrared transmission systems) can be swiftly deployed even in situations where "hard wired" networks are impossible or have been destroyed by disaster.

4. Computing trends are simple: faster, smaller, cheaper. Computers place powerful information systems in the hands of those who need them as tools in disaster response. Developments include:

■ High performance, battery-operated notebook computers that can be carried into the field.

■ Smaller (500 grams) "personal digital assistants" with the power of desktop machines of a decade ago and capable of being linked to networks, radio systems and GPS.

■ Robust, low-cost, portable data storage and distributions systems such as CD-ROM and data cartridges can put large amounts of data into the hands of users in the field.

■ Powerful PC-based database systems, including new applications using geographic information systems (GIS).

■ When combined with portable computers, simple data recording tools, such as "bar coding", offer new possibilities for data capture in the stressful environment of a disaster.

■ Beneath the "high tech" or "futuristic" image of telecommunications and information technologies, there is an increasingly robust, resilient and flexible suite of devices and services that have great relevance to disaster relief and provide appropriate and cost-effective tools.

publishing, and was used by peace and human rights groups, humanitarian agencies, journalists and academics.

And for the future...

In products and services the proportion of "software" – information, design, intelligence – rather than "hardware" – materials, manufacturing – is rising. The same process is underway in aid. In the past, lacking information or effective communications, agencies in effect had to act almost without thinking.

Today, under many growing pressures, from rising disaster complexity to falling donor funding, relief agencies have begun to think by using information in a tactical way, for example to better understand each situation or to prioritize their use of resources. While welcome, its tactical nature is seen in the enthusiasm to use technology to do the same as before but faster, rather than to explore what might now be possible that was once impossible.

One positive element has been growing discussion of using low-bulk, high-information relief in a catalytic way, and working with the existing culture and within the constraints of long-term complex conflicts from Afghanistan to Liberia. At the same time, in part to avoid what one journalist covering Rwanda dubbed "compassion without understanding", others have counselled caution about the way agencies work, the potentially-negative impact of aid and the need for far better understanding of culture, history, politics and community capacities.

From the latter perspective has come research by the US-based Collaborative for Development Action, published under the title *Do No Harm* – going back to the first principles of the medical profession's Hippocratic Oath – while the former idea, dubbed "smart aid" or "smart relief", has been best articulated by Professor Paul Richards, of the Joint War-Peace Research Programme, University College London and Wageningen University in the Netherlands (see Box 1.3).

Both make fundamental criticisms of present aid practice, most directly the warning from Richards that smart relief is essential because today's "bulk aid" misunderstands what is happening in many crises and can easily do tremendous harm by offering resources to fight over, sustaining belligerents and emphasizing the entitlement deficits of one side or another.

Richards has written that the need is for knowledge-intensive relief that can help stimulate cultural processes, such as radio. Unlike bulk aid, radio avoids creating patterns of patronage and clientship, which threaten some groups with social exclusion.

"Bulk relief may add to the tensions that sustain war because rationing decisions are made with only flimsy details about personal circumstances...provided there is open access for a range of citizen groups, and good feedback from rebel listeners, radio is capable of reaching the largest possible section of society without differentiation. As a mass medium, it signals incorporation rather than exclusion."

Smart aid envisages a level of political and practical understanding about the crisis almost impossible without indigenous leadership of the relief programme. As well as radio, among the factors it suggests should be fostered are "attack trade" commerce across conflict lines, distribution of the types of seeds crucial for crisis-coping strategies when careful cultivation is impossible, and even cultural movements, such as the inter-ethnic annual masquerades in Sierra Leone, or "investing in local `strolling players' to refresh the imaginations of war-zone minds temporarily stalled on a surfeit of hazard and horror".

This approach is not so new for some humanitarian agencies; even first aid implies avoiding doing harm through ignorance and applying cost-effective intelligence to a situation by spreading knowledge in communities, whether in the North or the South. Instead of an essentially one-way flow of information into agencies for disaster management, a two-way sharing of knowledge between agencies and disaster communities can now be implemented with growing ease through new information and communications technologies.

For example, instead of a million or more refugees spending ten years in neighbouring countries without much education beyond the basic, a very small investment in technology could offer knowledge transfer in subjects as diverse as health, agriculture or languages, and allow a significant growth in dialogue between beneficiaries and agency decision-makers.

Since disaster survivors are always the most important people in saving themselves and their neighbours, empowering the vulnerable before, during and after disasters has obvious benefits. Poor women are the most vulnerable people in the world; each year of a girl's education reduces the size of her future family and increases her long-term wealth. Targeting is easier than it might seem since the "who" of disasters is far clearer than the "where" or "when"; recurring disasters mean today's victim is likely to be tomorrow's victim, kept alive but lacking any new tools or knowledge to prevent or ease a future crisis.

> *...knowledge might be the best "product" for Northern NGOs to offer.*

Aid agencies do not need to invent all the skills and contents of the knowledge business: distance learning, among other skills, has become an enormous global industry, and is growing fast in the developing world through radio and internet. Net-capable hardware is cheap, especially as very simple DOS-based web browsers requiring far less processor power or memory are being developed. Global networks of low-orbiting satellites are being planned.

Digitally stateless

Reflecting on the impact of such changes on the so-called information poor, Professor Nicholas Negroponte, of the Media Lab at Massachusetts Institute of Technology, declared himself "very optimistic, especially about the developing world rapidly `becoming digital'". In helping those who are not just digitally homeless but digitally stateless, aid agencies have two clear roles in fostering knowledge transfer to vulnerable people: developing or adapting appropriate channels and materials, and encouraging Northern governments to pay for it.

There is plenty of work to be done. Even in what might be perceived as the simple subject of first aid, many different curricula are taught worldwide. The knowledge that vulnerable communities need varies enormously, from early warning systems to seed supplies. Delivery opportunities may be as short as a radio item to help listeners threatened by conflict or as long as years sitting in refugee camps, as varied as an elderly woman adapting to displacement in Europe and a child growing up in poverty on the cyclone-prone Bangladesh coast.

One way to maximize the value and sustainability of the "information aid" that a "knowledge agency" could offer would be to work closely with communities and their organizations so that vulnerable people help design it. Another way would be to create systems that make vulnerable people and local agencies the purchasers of such knowledge.

This might happen through simply offering plenty of choice to indigenous partners. More fundamentally, as Northern donors cut aid, they may prefer to put what money remains directly into the fast-growing new generation of dynamic and entrepreneurial NGOs in the South, and let them decide what they need from Northern NGOs. Given that almost all disaster materials and the majority of skills can be found within a country or nearby, knowledge might be the best "product" for Northern NGOs to offer.

The long-term future is obviously to use information and communications strategically, in ways which recognize the necessity for a clear shift of decision-making South, into the hands of indigenous disaster professionals and vulnerable communities. Such communities are by definition closer to their own disasters and in the best possible position – given the right information tools – to understand, prepare for, relieve and perhaps even prevent catastrophes.

Building global knowledge agencies that move information but leave the specifics of managing people and supplies to partner organizations on the spot offers a strategic approach to information and communications in disaster response, and would gradually allow a real aid partnership through a fundamental transfer of power into the developing world and towards vulnerable people themselves.

Chapter 3 *Sources, references and further information*

Anderson, Mary. *Do No Harm. Supporting Local Capacities for Peace through Aid.* Cambridge, Massachusetts: The Collaborative for Development Action, 1996.

Barling-Twigg, Jonathan. *Power Shift.* Information Strategy, Vol. 2 No. 1, London, 1997.

Borton, John. *Cooperation in the Sharing of Information for Emergencies.* United Nations Emergencies Unit for Ethiopia. Paper given at the UNDHA/ReliefWeb Workshop, October, 1996.

Davenport, Thomas H. *Process Innovation: Re-engineering work through information technology.* Boston: Harvard Business School Press, 1993.

International Federation of Red Cross and Red Crescent Societies. *Handbook for Delegates.* Geneva: 1996.

Negroponte, Nicholas. *The Next Billion Users.* Wired, Vol. 4 No. 6, New York.

Negroponte, Nicholas. *Being Digital.* New York: Knopf, 1995.

Richards, Paul. *Small Wars and Smart Relief; Radio and Local Conciliation in Sierra Leone.* In the Creative Radio for Development Conference Report, Health Unlimited, London, 1996.

Richards, Paul. *Fighting for the Rainforest; War, Youth & Resources in Sierra Leone.* London: James Currey, 1996.

Web sites

International Federation: http://www.ifrc.org

ReliefWeb: http://www.reliefweb.int/

Volunteers in Technical Assistance: http://www.vita.org

Wired: http://www.hotwired.com/wired/4.06/negroponte.html

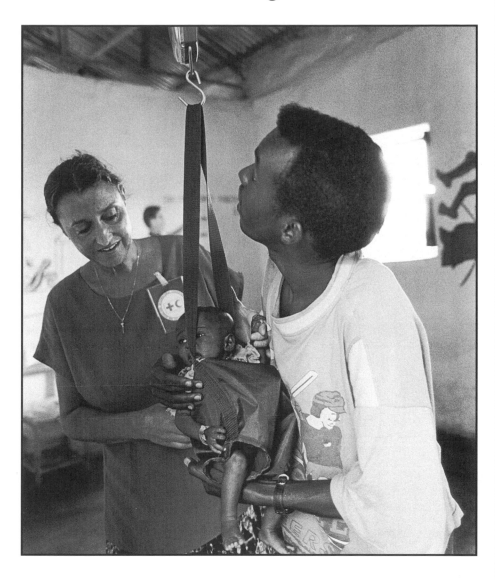

*Representative,
reliable and
accurate
information is the
fuel that drives
professional
humanitarian
operations. Data
determines who
is in need. Data
determines how
much is needed.
Data dictates
what is needed.*
Somalia, 1994.

Chapter

4 Epidemiological data for disaster decision-making

The thousands of Cambodian refugees who walked into eastern Thailand in 1979 were heading for what the media called "death camps". Escaping the war in Cambodia, they were exhausted, short of food, injured and heavily infected with malaria. Their high death rate was visible to the world when bodies were collected each morning for burial. When the massive relief operation began, no information was available to decide priority targets: children or adults, malnutrition or malaria, immunization, injuries or epidemic diseases. Without epidemiological data, the media reported a relief failure.

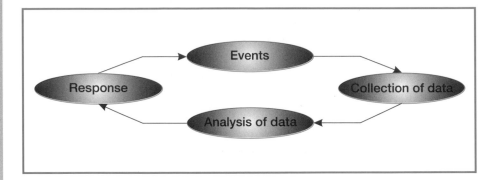

Public health surveillance activities were soon started, aiming to rapidly identify preventable causes of death, decide relief priorities, and monitor mortality and morbidity to see whether the relief effort was having an impact. The surveillance soon provided death rates, identified cerebral malaria as the principal cause of death and serious illness, and led to specific strategies for aggressive treatment. The swift decline in mortality during the next few weeks was directly linked to a relief effort that correctly targeted the main preventable causes of death. Collection of simple epidemiological data, "quick and dirty" field surveys targeted to specific relief questions, and a brief weekly surveillance report made the relief effort responsive to priority health needs and provided reliable information for donor organizations and the media.

As the "death camps" proved, better epidemiological knowledge of the causes of death, and types of injuries and disease caused by disasters, is essential to determine what is required – supplies, equipment, personnel – for effective response. Proper decision-making in relief management requires timely and appropriate information; in several important areas epidemiology can help make disaster management more effective, and there are a number of potential improvements that would save lives and cut costs. In recent disasters, epidemiologists have been able to define quickly the nature and extent of health problems, assess medical needs, match available resources to those needs, identify groups at particular risk of death, injury and illness, monitor relief effectiveness and provide recommendations to decrease the consequences of future disasters.

Epidemiology is the quantitative study of the distribution and cause of diseases in human populations; epidemiologists look at populations, not individual patients. Central to epidemiology is the belief that diseases do not occur at random, but have predictable patterns that can be studied and used to develop prevention strategies. Such patterns may be clusters of disease, injuries or other health outcomes in time, space or among groups of people. Epidemiological principles have made significant contributions to the understanding of diseases and disease transmission, even when the causative agent is not known. Long before the AIDS virus was discovered, epidemiological studies had identified high-risk groups, risk factors for development of the disease, means of transmission, and had suggested ways to prevent its spread.

Applying epidemiology in disasters must be seen in a broader perspective, linking data collection and analysis to immediate decision-making. In recent years, epidemiological techniques have become a key component in many disaster relief operations.

Practical application of epidemiology to disaster management began in the late 1960s with crises in Nigeria, the West African Sahel, Ethiopia, Bangladesh and Uganda. Techniques were developed by the Centers for Disease Control (CDC), the

Centre for Research on the Epidemiology of Disasters (CRED) and the World Health Organization (WHO) to rapidly assess nutritional status and identify populations in need of medical and public health assistance.

Epidemiological surveillance in disaster situations aims to collect, analyse and respond to data. The surveillance cycle must turn many times: immediately, with rapid, "quick and dirty" assessments of problems using rudimentary data-collection techniques; then with short-term assessments involving the establishment of simple but reliable sources of data; and, later, using ongoing surveillance to identify continuing problems and monitor the impact of interventions.

Pervasive cycle

Several past disaster-relief operations have shown the need for better epidemiological information. Aid staff have been forced to mount major relief efforts with no reliable information on the health of the affected population. Without adequate data, the response was often dictated by the assistance available from donors or was based on unverified reports, myths and stereotyped forms of aid. Disasters were often cluttered by unnecessary, outdated or unlabelled drugs, vaccines that were ineffective, not needed or improperly used, medical and surgical teams without proper logistical support, and relief programmes that did not address immediate needs.

The media often called relief, with its medical components, "the second disaster". Although many governments and volunteer agencies have developed extensive disaster-response capabilities, under-use and lack of coordination of disaster assessment have contributed to a pervasive cycle of inappropriate, and often ineffective, disaster relief.

After the 1976 Guatemala earthquake, 100 tonnes of unsorted medicines were airlifted to the country from international donors, 90 per cent of which were of no value since they had expired, were already opened or were labelled in foreign languages. More than a decade later, after the 1988 Armenian earthquake, relief organizations sent at least 5,000 tonnes of drugs and consumable medical supplies, of which – because of identification and sorting difficulties – 11 per cent were useless, another 8 per cent had expired and only 30 per cent were immediately usable. Ultimately, 20 per cent of all the drugs provided by international aid had to be destroyed.

Data collection should be limited to information that can and will be acted upon.

Critical to any disaster response in determining relief priorities is an early and effective damage assessment to identify urgent present, and possible future, needs of the affected population. Disaster assessment from rapid field investigations provides relief managers with objective information about the effects of the disaster on a population, allowing maximum efficiency by matching available resources to emergency needs. Fast assessment and mobilization of the right resources can significantly reduce death and illness caused by a disaster.

Data-collection techniques (primarily random or "convenience" samples, systematic surveys and simple reporting systems) are methodologically straightforward and, if suitable staff and transport are available, they should provide rapid and reasonably-accurate estimates of relief needs. Elaborate sampling techniques are not needed; in these situations, the adage "being roughly right is generally more useful than being precisely wrong" applies. A balance exists between ensuring that survey procedures are simple yet effective and achieving a high degree of validity and reliability.

Data collection should be limited to information that can and will be acted upon. Since complex reporting requirements will not be followed, information that is not immediately useful should not be collected during the emergency phase of a relief operation. The most valuable data are generally simple to collect and analyse. Standard clinical case definitions for the most common causes of illness and death should be developed and put in writing.

Data collected in an emergency health assessment usually fall in one of the following categories:

■ overall magnitude of disaster impact (geographical extent, number of affected people, estimated duration);

■ impact on health (number of casualties);

■ integrity of health delivery systems;

■ specific health needs of survivors;

■ disruption of "lifelines" and other services (power, water, sanitation, housing) relevant to public health; and

■ extent of disaster response by local authorities.

Figure 4.2 Characteristics of data-collection methods in disaster settings

ASSESSMENT METHOD	TIME REQUIRED	RESOURCES	DATA-GATHERING TECHNIQUES + INDICATORS	ADVANTAGES	DISADVANTAGES
1 **Predisaster background data**	Ongoing	Trained staff	Reporting from health facilities and practitioners + Disease patterns and seasonability	Provides baseline data for detecting problems and assessing trends	None
2 **Remote: planes, helicopter, satellite**	Minutes/hours	Hardware	Direct observation, cameras + Destroyed structures, flooding	Quick; useful when ground transport out or to identify area affected	Expensive; large objective error; minimal specific data
3 **On-site "walk/ride through"**	Hours/days	Transport, maps	Direct observation, talks with local leaders and health workers + Deaths, homeless people, numbers and types of diseases	Quick; visible; does not require technical (health) background	Non-quantitative data; risk of bias; high error rate; "worst" areas may be unreachable
4 **"Quick and dirty" surveys**	2-3 days	Few trained staff	Rapid surveys + Deaths, no. hospitalized, nutritional status,(see also 3)	Rapid quantitative data; may prevent mismanagement; can provide data for surveillance	Not always random samples; labour intensive; risk of overinterpretation

Adapted from "Nieburg's Model for Data Collection Methods in Disaster Situations," in *Health Aspects and Relief Management After Natural Disasters*. Centre for Research on the Epidemiology of Disasters. Brussels, Belgium, 1980

Such information should be used to plan and implement immediate responses. The emphasis of these assessments is to gather a small amount of relevant information quickly (usually hours to a few days after a sudden-impact disaster). A "quick and dirty" approach may require a multidisciplinary team (e.g., epidemiologist, statistician, engineer and health planner, among others) and relies on visual inspection (e.g., aerial fly-overs), interviews with key personnel and field surveys.

The health effects of a disaster can be measured by indicators that permit objective assessment to guide relief efforts. Such indicators following an earthquake, for example, include mortality; morbidity; numbers of damaged houses, homeless people and non-functioning hospitals; and the status of lifelines (e.g., water, electricity, gas, sewage disposal). Since everyone in a disaster area has needs and losses, the challenge of early assessment is to decide priorities: How can swift intervention save the most lives or prevent the greatest illness?

Early assessments should address basic questions. Later surveys can examine more carefully the availability of medical care, the need for specific interventions and

---

Box 4.1 **Hurricane Andrew, cluster surveys and health coordination**

When Hurricane Andrew struck south Florida in August 1992, epidemiologists demonstrated the use of a modified cluster-sampling method to perform a rapid needs assessment. In the first survey, three days after the hurricane, clusters were systematically selected from a heavily-damaged area by using a grid overlaid on aerial photographs. Survey teams interviewed seven occupied households in consecutive order in each selected cluster. Results were available within 24 hours of beginning the survey. Surveys of the same heavily-damaged area and of a less severely-affected area were conducted seven and ten days later, respectively.

The initial survey found few households with injured residents, but many without telephones or electricity. These findings convinced disaster-relief workers to focus on providing primary care and preventive services rather than to divert resources towards unnecessary mass-casualty trauma services.

The cluster-survey method used was modified from methods developed by the WHO's Expanded Programme on Immunization (EPI) to assess vaccine coverage. Although cluster surveys have been used in refugee settings to assess nutritional and health status, this represented the first use of the EPI survey method to obtain population-based data after a sudden impact natural disaster.

In the hurricane, medical systems suffered severe damage. Acute-care facilities and community health centres were closed and doctors' offices destroyed. State and federal public health officials, the American Red Cross and the military established temporary medical facilities. In the four weeks after the hurricane, officials established disease surveillance at 15 civilian and 28 military free care centres, and at eight emergency departments in and around the impact area. Public health workers reviewed medical logbooks and patient records daily, and tabulated the number of visits using simple diagnostic categories (e.g., diarrhoea, cough, rash).

The surveillance was able to characterize the health status of the hurricane-affected population and to evaluate the effectiveness of emergency public health measures. Data from the system indicated that injuries were an important cause of morbidity among civilians and military personnel, but that most injuries were minor. Surveillance information was particularly useful in responding to rumours about epidemics, so avoiding widespread use of typhoid vaccine, and in showing that large numbers of volunteer health care providers were not needed.

Although the surveillance achieved its objectives, there were several problems. First, relief agencies needed to coordinate their efforts. Data from the civilian and military systems had to be analysed separately because different case definitions and data-collection methods were used. Second, there was no baseline information available to determine whether health events were occurring more frequently than expected. Third, rates of illness and injury could not be determined for civilians because the size of the population at risk was unknown.

Although proportional morbidity (number of visits for each cause divided by the total number of visits) can be easily obtained, it is often difficult to interpret. An increase in one category (e.g., respiratory illness) may result from a decline in another category (e.g., injuries), rather than from a true increase in the incidence of respiratory illness.

---

*Figure 4.3
Crude death or
mortality rates
(CDR or CMR) are
the most critical
indicators of a
population's
improving or
deteriorating
health status and
are the category
of data to which
donors and relief
agencies most
readily respond.*

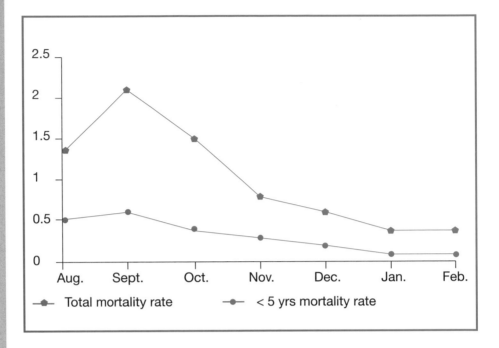

epidemic control. Adequate health-related data during the initial assessment will often serve as the basis for an epidemiological surveillance system. Such surveillance must monitor the impact of relief on health problems, to determine whether relief efforts are having a tangible impact or if new strategies are needed. Surveillance becomes a cyclical process, constantly monitoring and assessing simple health interventions and their outcomes. Sustained feedback about changing needs allows disaster response to be modified.

Using standard case definitions and data-collection protocols, surveillance can be implemented in several settings, from existing hospitals to disaster-assistance centres. After the 1988 floods in Khartoum, Sudan, medical clinics in each of three urban districts collected surveillance data. Clinics were chosen on the basis of their accessibility to surveillance workers and because their patients were presumed to be representative of the flood-affected population. Data obtained showed that diarrhoeal disease accounted for the greatest number of clinic visits, followed by malaria.

> *Data must be collected rapidly under highly-adverse conditions.*

Despite widespread acceptance of the need for disaster intelligence, it is extremely difficult to collect accurate data immediately after an emergency. Few schools of medicine or public health teach the applications of epidemiology, and most health ministry officials are not aware that epidemiology is a useful management tool in emergencies. Local authorities or relief agencies may be reluctant to commit scarce resources to assessment – such as aerial surveys – at the expense of other operational activities. Data must be collected rapidly under highly-adverse conditions. Transport, access and communications may be difficult or impossible. The area may lack indigenous epidemiological expertise to conduct rapid assessment surveys, while – assuming local authorities do not object to external assessment teams – the unfamiliarity of outside experts with local langua-ges and customs could compromise data quality. The lack of good information is compounded if the local or international media offer inaccurate information.

The effectiveness of emergency health-information management following a disaster can be measured by how rapidly data collected and analysed can identify prevention strategies and how effectively these strategies can be implemented by decision-makers. This requires active coordination between epidemiologists and decision-makers, who must understand the data and strategies, and implement the required policies to direct relief and cut the death rate.

Complex emergencies

One measure of the technical progress in managing complex humanitarian emergencies during the past 20 years is the almost routine establishment of health information systems to monitor refugee populations. Particularly during the past five years, humanitarian organizations such as the United Nations High Commissioner for Refugees (UNHCR) and Médecins sans Frontières (MSF) have recognized the importance of a standard health information system to track disease patterns and to assist health planning and evaluation. Such agencies routinely call on epidemiologists to conduct surveys during the emergency phase and then set up surveillance systems to assess both programme effectiveness and the need for further public health interventions.

Data collected in refugee camps generally fall into these categories:

- total affected population;
- age-sex breakdown;
- identification of at-risk groups, e.g., children under five, pregnant and lactating women, disabled and wounded people, unaccompanied minors;
- average family or household size;
- mortality;
- morbidity;
- nutritional status; and
- health programme activities.

Graveyard watch

Total population figures are needed to calculate quantities of relief supplies and as the denominator for all birth, death, injury, morbidity and malnutrition rates. Rates permit valid comparisons of morbidity and mortality between populations that differ in size and composition (e.g., age, race or sex). Population breakdown by age and sex allows for calculation of age- and sex-specific rates within vulnerable groups, and enables interventions (e.g., immunization campaigns) to be targeted effectively and shifted over time, so that an early focus on under-fives, for example, can be widened if the situation of other groups worsens. Data on arrivals and departures can help predict future needs and influence long-term interventions, such as tuberculosis therapy.

Mortality surveillance is vital and may require creative methods, such as a 24-hour graveyard watch or monitoring of burial-shroud distribution, that illiterate disaster victims or aid staff can manage or support. During the acute phase of relief operations for Ethiopian refugees in both Somalia in 1980 and Sudan in 1985, death rates were 18 to 45 times greater than the death rates of surrounding non-refugee populations. During the 1992-1993 famine in Somalia, mortality rates for displaced people in temporary camps were more than twice as high as rates for settled persons.

While mortality rates among young children are usually higher even in normal situations, in emergencies children can be at exceptionally high risk, especially if unaccompanied or orphaned. Death rates among children under five in Sudanese camps were almost eight times higher than death rates among people aged 15 or older. In Somalia, mortality rates for these children were two to four times greater than older people in the same communities. During 1985, 9 per cent of the Tigrayan refugees in the camps of eastern Sudan died, including – based on the population's age structure – 10,000 to 15,000 children. Surveys in Sarajevo determined that the 1992 death rate was four times the pre-war rate; most excess mortality was due to war-related trauma. Death records in the eastern Bosnian enclave of Zepa, where the pre-war population of 7,000 increased to 33,000 as a result of an influx of displaced Muslims, show the same trend.

Surveillance of nutritional status and diseases of public health importance, from measles to meningitis, will help a range of actions: planning refugee preventive and curative health programmes; procurement of appropriate supplies; recruitment and training of medical personnel; and improving environmental sanitation (e.g., toward mosquito control in areas where malaria is endemic). A range of disaster victims and local aid staff, from traditional birth attendants to community health workers, can be trained to recognize the clinical signs and symptoms of common diseases and conditions in refugee camps. Prevalence of acute malnutrition acts as an indicator of the adequacy of the relief ration. Widespread malnutrition, despite an adequate average daily ration, may indicate inequities in food distribution or high rates of communicable diseases (e.g., diarrhoea), while the presence of nutritional deficiency disorders (i.e., pellagra, anaemia or xerophthalmia) indicates the need for ration supplements.

Surveillance systems among Rwandans in eastern Zaire from 1994 to 1996 were an excellent illustration of recent progress in refugee camp information management. Data collection assisted identification of disease outbreaks, implementation and assessment of interventions (e.g., diarrhoeal disease control through provision of

---

Box 4.2 **Monitoring mortality among refugees in eastern Zaire**

Crude death or mortality rates (CDR or CMR) are the most critical indicators of a population's health status and are the category of data to which donors and relief agencies most readily respond. A CMR also provides a baseline to judge the effectiveness of relief programmes.

During the emergency phase of a relief operation, death rates should be expressed as deaths/10,000/day to allow for detection of sudden changes. In general, health workers should be extremely concerned when CMRs in a displaced population exceed 1/10,000/day, or when under-five mortality rates exceed 4/10,000/day.

In eastern Zaire in July 1994, the CMR among one million Rwandan refugees ranged from 34.1 to 54.5 deaths per 10,000 per day, among the highest ever recorded. Between 6 per cent and 10 per cent of the refugee population died during the month after arrival in Zaire, a death rate two to three times the highest previously-reported rates among refugees in

Thailand (1979), Somalia (1980) and Sudan (1985). This high mortality was due almost entirely to the epidemic of diarrhoeal diseases.

By the third week of the refugee influx, relief efforts began to have a significant impact. Routine measures, such as measles immunization, vitamin A supplements, standard disease treatment protocols and community outreach programmes, were established in each camp, and the water distribution system began to provide an average of 5 to 10 litres per person per day.

A consensus was quickly reached among multiple agencies and organizations on standard information to collect, and a high level of cooperation and coordination of public health programmes was achieved. The relief programme in the Goma camps, based on rapidly-acquired health data and effective interventions, was associated with a steep decline in death rates to five to eight per 10,000 per day by the second month of the crisis.

clean water and sanitation, soap distribution, and aggressive rehydration therapy), and recognition of the need for more primary health-care services. The experiences in Goma underscore the need for simple data-collection efforts and for the selective targeting of surveillance efforts toward the most important health and medical problems (e.g., watery and bloody diarrhoea) during the emergency phase in refugee camps.

Food rations

It is essential to integrate all components of an emergency health information system. Analysis of a single element may be misinterpreted. An apparent decrease in malnutrition prevalence, for example, should be interpreted in the context of childhood mortality rates. In the 1991 Kurdish refugee crisis, the overall prevalence of wasting was misleadingly low, since many of the youngest and most severely-affected children had already died.

Information on programme coverage and effectiveness of interventions should also be systematically collected; such data should include the average quantity of food rations distributed, the availability of clean water per person, the ratio of families to latrines, immunization coverage and supplementary-feeding programme attendance. Tracking health sector activities will help determine whether certain groups are under-served.

Complex humanitarian emergencies lack mechanisms to evaluate the effectiveness and efficacy of health response and outcome. There are five basic categories of data (indicators or measures of effectiveness) useful for monitoring and evaluating emergency-relief programmes:

1. Mortality:

■ monthly mortality rate;

■ under-five mortality rate.

2. Morbidity:

■ serious communicable diseases (cases/month);

■ nutrition-related diseases (protein energy malnutrition, anaemia, scurvy, pellagra, etc.).

3. Nutritional status:

■ screening of new arrivals by mid-upper arm circumference (MUAC);

■ periodic surveys by weight-for-height.

4. Activities:

■ immunizations;

■ feeding centre attendance;

■ other public health activities;

■ numbers of outpatients, inpatient admissions, referrals.

5. Vital sectors:

■ food distributions, rations;

■ water, sanitation;

■ shelter, blankets, clothing;

■ domestic utensils, cooking fuel.

Figure 4.4
Data on diseases
of public health
importance may
help plan an
effective
preventive and
curative health
programme for
refugees.

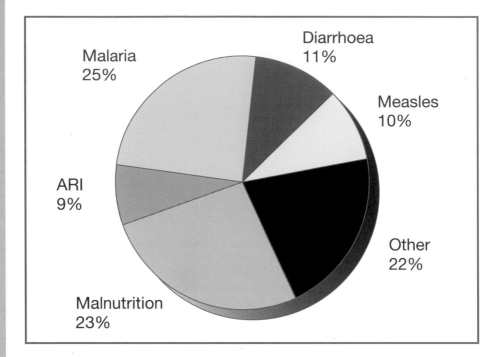

Improvement rates in camp populations have varied considerably. Mortality rates fell rapidly in Cambodian refugee camps in Thailand in 1979-1980 and in Kurdish camps in Turkey in 1991. Improvement in health and nutritional status was much slower among recent Sudanese and Somali refugees in camps in remote areas of Ethiopia and Kenya respectively, where water supplies were completely inadequate and food delivery logistics were especially difficult.

Overall, less than 1 per cent of Cambodian refugees in Thai camps died during the first 12 months; in contrast, 9 per cent of the refugees in eastern Sudan died during the same period. In eastern Ethiopia in 1988-1989, mortality rates among Somali refugees were initially low, but increased after six months and peaked after nine months. Paralleling the increase in mortality rates, malnutrition prevalence rates among children under five years' old increased from approximately 15 per cent to more than 25 per cent between March and May 1989, more than six months after the international agencies launched the assistance programme. Inadequate general food rations probably caused this deterioration in nutritional status.

As the *World Disasters Report 1996* (Chapter 3, Nutrition and food aid) outlined, the use of health information to guide decision-making will be helped if critical indicators are established from the start of a relief operation. For example, a measles incidence rate of 1/1,000/month might be an indicator that would initiate specific preventive actions. Similarly, during a cholera outbreak, a case fatality rate of 3 per cent in a given week might stimulate a review of treatment procedures.

Aiding its evolution as needs change, the health information system should be periodically assessed to determine accuracy, completeness, simplicity, flexibility and timeliness. Data should be presented to decision-makers in simple, clear tables and graphs, and their use of the data assessed. Most importantly, there should be regular feedback to data providers in the field through newsletters, bulletins and frequent supervisory visits.

Box 4.3 Epidemiological implications from the siege of Sarajevo

Epidemiological surveillance during the siege of Sarajevo showed a range of specific impacts, one clear aspect of which was the effect on women, with strong implications for future relief programmes in similar situations.

Food shortages, poor water supplies, limited fuel and the fighting itself all show up – directly and indirectly – in figures for reproductive health. Shelling destroyed the central hospital's obstetrics and gynaecology facility, cutting bed numbers from 450 to 50 and operating rooms from four to one, sniping discouraged attendance at antenatal clinics, outbreaks of hepatitis and diarrhoeal diseases were common, and dietary deficiencies increased.

Amid a shifting population, with many men away fighting, and fear of the future discouraging pregnancy, annual live births fell from 10,000 pre-war to 2,000. Contraceptive use stayed low, at about 5 per cent, but abortion increased significantly to two terminations for every pregnancy taken to term.

Although maternal mortality did not grow, perinatal mortality rate rose from 15.3 per 1,000 live births to 38.6. Major causes of perinatal mortality were respiratory distress syndrome, congenital abnormalities and asphyxia, problems that in many cases could have been more easily tackled in facilities under normal conditions.

Causes of the steep rise in low birth weight (<2,500g) from 5.3 per 1,000 live births to 12.8 could have included stress, far more active and passive smoking, especially among women, and lower food intake. As the siege began, early weight loss among adults averaged 10 to 14kg. But a rise of congenital abnormalities involving hydrocephalus or anencephalus or both, from 0.37 per cent to 3 per cent, suggests serious deficiencies in specific nutrients.

The implications of these impacts upon reproductive health include concern at the future well-being of many infants, especially those of low birth weight, and fears about the potential long-term negative effects on fertility and mental health.

Where the health system is deliberately targeted to undermine the security of the population, international aid should clearly aim to at least maintain, if not improve, health services, while the evidence of Sarajevo suggests a failure to meet the fundamental needs of women, from contraception to food supplements.

The full potential of surveillance information in guiding relief efforts and developing preventive measures has not been realized. Decision-makers must develop a broader appreciation for the types of surveillance data and its range of uses. Those who develop the system must demonstrate that its benefits – having data to respond to community concerns – outweigh its costs, and recognize that disseminating information is as important as collecting and analysing data.

The capacity to conduct surveillance must be supported by training personnel at all levels of the health sector in the practice and application of epidemiology and surveillance. Gaps in surveillance methodology and data coverage must be closed. Surveillance systems must be subjected to rigorous evaluation to ensure that they meet their established objectives.

Knowledge gaps

Relatively few surveillance efforts have been evaluated in a systematic fashion. Identifying successful data-collection strategies and acknowledging shortcomings will yield critical information for developing better methods of surveillance for the next disaster.

Widely-accepted, standard case definitions for disaster-related deaths or injuries do not exist. As a first step, easily-understood case definitions of disaster-related illnesses and injuries should be developed, agreed upon by relief agencies, and disseminated to public health and emergency management institutions. Reporting procedures and standard forms that can be modified for use in many settings should be developed. Allowance should be made for gathering sufficiently-detailed information, such as the cause and circumstances of certain injuries, to guide interventions and prevention efforts. Case definitions and standardized reporting procedures will help expand surveillance activities after a disaster to hospital emergency departments, outpatient medical clinics, first-aid shelters, and doctors' offices.

> *Collecting more information will not automatically improve disaster response.*

Without baseline information on the population at risk, epidemiologists must use indirect measures, such as proportional morbidity, to estimate risk. Public health officials should consider collecting baseline information, such as distribution of types of illnesses and injuries at various health facilities, as an integral component of disaster preparedness. Another useful element of baseline information is beneficiary capabilities, so that arriving relief agencies can create systems that make maximum use of existing skills within the disaster-affected community itself.

Finally, no standard methods or indicators exist to determine rapidly the needs of disaster victims and communities. Assessment indicators and surveillance methods in disaster settings should possess all the four attributes crucial to disaster relief: simple to use, timely, collectable under adverse field conditions, and useful.

The following actions would improve both data collection and use:

■ Develop and disseminate standard case definitions of disaster-related morbidity (illnesses and injuries) and mortality.

■ Develop standard reporting forms and procedures that can be easily modified for use in various settings. *Epi Info*, a public-domain software package developed by the CDC, can be used to generate questionnaires as well as to enter and analyse data using a laptop computer.

■ Use and modify sampling techniques, such as cluster sampling, for conducting rapid assessments and estimating health service needs.

■ Investigate the use of existing electronic data systems, such as hospital patient discharge data, as potential sources of timely information on morbidity.

■ Test the feasibility of establishing and maintaining networks of physicians and health care facilities, particularly in areas at high risk for recurrent disasters.

■ Explore the use of Internet and other forms of electronic communication for collecting and disseminating emergency surveillance data.

■ Make more use of existing information systems (e.g., national Red Cross or Red Crescent societies, national meteorological services, civil defence, insurance industry) and record linkages to establish databases suitable for epidemiological research on the health impact of disasters.

Summary

In disaster relief, important and often irreversible decisions must be made fast. The need for reliable early data to assist these decisions is crucial. An organized approach to data collection in disasters can greatly improve decision-making and can predict a variety of options for disaster-relief officials to face. Collecting more information will not automatically improve disaster response.

Much of the information presently generated by governments, international agencies, voluntary relief agencies and other groups is of questionable validity, and a lot of it is improperly collected, analysed and used. To save more lives and cut costs, standard procedures for collecting data in disasters must be developed that can then be linked to operational decisions and effective public health interventions.

Although during the past few years, epidemiologists are routinely called upon by relevant UN agencies and NGOs to conduct health and nutrition surveys following disasters, the failure of the modern international humanitarian assistance community to make the most effective use of existing epidemiological knowledge continues to cost many lives every year.

Chapter 4 *Sources, references and further information*

Autier, P. et al. *Drug supply in the aftermath of the 1988 Armenian earthquake.* The Lancet, vol. 335, pp.1388-1390, 1990.

Carballo, M. et al. *Health in countries torn by conflict: lessons from Sarajevo.* The Lancet, vol. 348, pp. 872-874.

Centers for Disease Control. *Famine-affected, refugee, and displaced populations: recommendations for public health issues.* MMWR, no. 41 (RR-13), pp. 1-76, 1992.

Glass, R.I. et al. *Rapid assessment of health status and preventive medicine needs of newly arrived Kampuchean refugees, Sa Kaeo, Thailand.* The Lancet, pp. 868-872, 1980.

Goma Epidemiology Group. *Public health impact of the Rwandan refugee crisis: what happened in Goma, Zaire in July, 1994.* The Lancet, vol. 345, pp. 339-344, 1995.

Guha-Sapir, D. and Lechat, M.F. *Information systems and needs assessment in natural disasters: an approach for better disaster relief management.* Disasters, no. 10, pp. 232-237, 1986.

Guha-Sapir, D. *Rapid needs assessment in mass emergencies: review of current concepts and methods.* World Health Statistics Quarterly, no. 44, pp. 171-181, 1991.

Internation Centre for Migration and Health. *Reproductive health and pregnancy outcome among displaced women.* Report of the Technical Working Group, October 1995, Geneva.

Lechat, M.F. *Updates: the epidemiology of health effects of disasters.* Epidemiological Review, no. 12, pp. 192-197, 1990.

Logue, J.N. et al. *Research issues and directions in the epidemiology of health effects of disasters.* Epidemiological Review, no. 3, pp. 140-162, 1981.

Noji, E.K. (ed.). *Public health consequences of disasters.* New York: Oxford University Press, 1997.

Seaman, J. *Epidemiology of natural disasters.* Contributions to Epidemiology and Biostatistics, no. 5, pp. 1-177, 1984.

Western, K. *The epidemiology of natural and man-made disasters: the present state of the art.* Dissertation, University of London, 1972.

Smillie, Ian and Helmich, Henny. *Non-Governmental Organizations and Governments: Stakeholders for Development.* Paris: OECD, 1993.

Web sites

Centers for Disease Control: http://www.cdc.gov/cdc.html

Center of Excellence in Disaster Management and Humanitarian Assistance: http://www.204.208.4.136/

Global Health Disaster Network: http://www.hypnos.m.ehime-u.ac.jp/GHDNet/index.html

Pan-American Health Organization: http://www.who.paho.org/

World Association for Disaster and Emergency Medicine: http://www.hypnos.m.ehime-u.ac.jp/GHDNet/WADEM/index.html

World Health Organization: http://www.who.ch/

Behind the headlines are the people whose capacities and resilience provide the base for humanitarian action. The aid agencies' commitment to improving standards is first and foremost a commitment to them, the beneficiaries

Georgia, 1996.

Chapter

5

Aid trends in 1996: raising questions amid falling funds

Until the dramatic events in the Great Lakes region at the end of the year, one might have been forgiven for thinking that all was quiet on the humanitarian "front" in 1996. The Dayton accords on Bosnia were promoted as the "beginning of the end" of the largest relief operation in Europe since the end of World War II and the refugee situation in the Great Lakes had reached an impasse with no obvious end in sight.

But for many aid agencies, 1996 was a year of intense reflection and learning following their experiences in Rwanda, Bosnia and Somalia. Publication of key evaluations of major operations in Rwanda, Sudan and Liberia raised hard questions about the rationale and organization of international responses to complex political crises. These reports added weight to discussions about fundamental principles of relief action, and whether and how these can be codified and monitored. The debate has broadened considerably from a focus on relief to a much broader one on humanitarian aid, with its notions of what constitutes assistance and its emphasis on the way in which relief is provided in conflict zones and the values that determine the way assistance is provided.

For many agencies, 1996 was also a difficult financial year as emergency-aid budgets continued their downward trend from the 1994 peak. Some agencies were obliged to downsize operations to match expenditures with income, making the reflection and learning more questioning and introspective.

> *....a few Northern states effectively decide the amount of aid and how it is distributed.*

Finally, the fighting in eastern Zaire, the return of many thousands of refugees, questions about how many refugees still remained in conflict areas and debates over the need for military intervention brought an increased interrogation by journalists of the role and accountability of non-governmental organizations (NGOs). If such criticism marks more than a temporary shift in the once mutually-supportive relationship between agencies and the media, the repercussions could be of long-term significance.

Financial trends

Analysis of trends in official development assistance (ODA), and of emergency relief aid in particular, provides a starting point from which to assess the state of the humanitarian system.

It is a tradition on these pages to preface discussion of trends in relief financing with a lament for the quality of data available. This year is no exception. The database maintained by the Development Assistance Committee (DAC) of the Organisation for Economic Co-operation and Development (OECD) remains the sole source of official data on relief expenditures.

This source is problematic as it does not differentiate "relief" food aid from "developmental" food aid and there are questions about the accuracy of categorization by individual donor organizations. The picture is becoming more blurred as donors establish new budget lines for rehabilitation assistance.

Over the past ten years the total value of the aid budget has fluctuated markedly (see Figure 5.1). Importantly, however, compared with previous decades, the value of ODA has fallen sharply relative to income of the donor countries. In mid-1996, aid provided by member states of the OECD had fallen to 0.27 per cent of GNP, the lowest recorded since the United Nations (UN) established its target of 0.7 per cent. Rich countries are increasingly unwilling to redistribute their wealth to the poorer ones.

Falling ODA from OECD members is even more important because their share of global ODA has risen over the past decade. In 1985 OECD donors supplied 87 per cent of total global ODA. Declining contributions from oil-producing states, particularly in the Middle East, left OECD members accounting for over 98 per cent by 1993. The policies of a few Northern states effectively decide the amount of aid and how it is distributed.

The proportion of aid resources being transferred to the poorest countries has fallen by nearly 10 per cent, while more aid is going to the former Soviet Union and Eastern

Figure 5.1
Total amounts of
official
development
assistance.

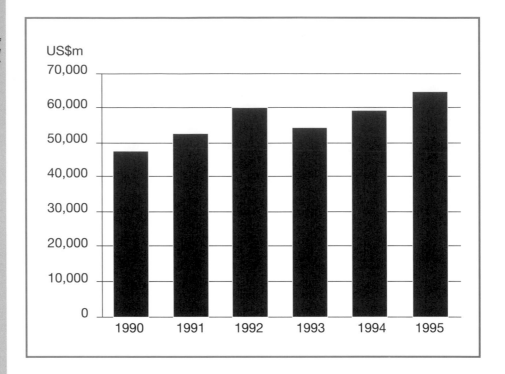

Europe. Peacekeeping costs are also rising: between 1990 and 1994 they more than doubled from 2.4 billion US dollars to $5.7 billion. While peacekeeping costs do not always come directly from aid budgets, donor governments often point to the increasing financial burden of such military operations when explaining aid cuts.

Until recently, it seemed emergency aid was relatively protected from declining ODA because it was favoured by politicians keen to present their humanitarian credentials on television. Even if that was the case, annual levels of emergency aid are highly event-specific. Massive expenditures for Gulf War population displacements, and relief operations in the former Yugoslavia and Somalia from 1991 to 1994 are evident in Figure 5.2, as is the impact of the spending in and around Rwanda during 1994. The most recent figures available show that distress relief expenditures fell marginally as a proportion of total ODA from 5.86 per cent in 1994 to 5.19 per cent in 1995 (excluding food aid). For some agencies, the fall has been more pronounced, forcing retrenchment.

These figures suggest that while relief expenditures may rise in response to particular phases of specific crises, actual relief receipts by conflict-affected populations remains relatively static and in some cases may actually decline. Despite the protracted nature of complex political emergencies, funding for the basic needs of conflict-affected populations does not appear to be sustained. Evidence from Sudan and Liberia revealed by recent evaluations indicates that cuts in relief funding, often justified on only the scantiest of evidence, are having a real, and negative, impact on the welfare and physical well-being of some of the most deprived people in the world. The onus is on relief agencies to defend populations' entitlements to at least a minimum of assistance, while ensuring that available funds are spent in ways which maximize impact.

Definitions used to disaggregate expenditure types are increasingly inadequate in capturing the real nature of the crises. The relatively-simplistic categories of "relief" and "development" assistance should be abandoned for categories of expenditure which reflect more accurately the persistent nature of the crisis for people suffering the effects of war and structural economic decline. Such a breakdown, similar to

---

Box 5.1 `Missing' Rwandan refugees; from analysis to advocacy

The return of some 600,000 refugees to Rwanda from Zaire in November 1996 prompted an extraordinary debate about the numbers still in Zaire, their condition, whether military intervention was required, the role of media and how much aid agencies could be trusted.

The immediate media interpretation was of a story with a moderately happy ending – most if not all refugees in eastern Zaire had returned in fairly good spirits and health, while the extremists who had forced them to remain in exile flee west, deeper into Zaire. Reality was more complicated. The return flow was only about half of the 1.1 million refugees agencies said they had been helping since 1994. Opinion was divided about the remaining 500,000.

The Rwanda government said that the original refugee figure had been grossly over-inflated and that all refugees had returned, a view broadly supported by the United States despite American aerial reconnaissance photographs showing large groups of people on the move in eastern Zaire. Led by UNHCR, aid agencies argued that the "lost" refugees were heading west ahead of advancing rebel troops. Estimates of numbers varied widely.

Events have since proved the aid agencies largely correct, with hundreds of thousands of refugees moving away from the border and ahead of rebel forces. At the time, doubts were strong enough to help shelve plans for a multinational, military-intervention force, which had originally been intended to open the way for the return of all Rwandan refugees from Zaire.

Scepticism about numbers quoted by aid agencies was bolstered by some journalists and aid observers, who argued that the agencies had a vested interest in overstating the numbers and underestimating their resources and capacity to survive. These commentators said, perhaps with reason, that aid agencies had in past "cried wolf" too often and that for the unscrupulous there was a constant imperative to raise funds through images of lost and hungry refugees.

Aid agency staff, while accepting that some of those on the move were armed militia who might well be guilty of genocide, warned that branding every Hutu an extremist and allowing the rough justice of

a rebel war to take its course – including the deaths of women and children – should not be acceptable policy for either agencies or governments.

The credibility of aid agencies was already suspect in some quarters because some had overemphasized the danger of hunger and disease when refugees were trapped in camps around Goma as rebels seized the town. After television viewers saw refugees returning to Rwanda in fairly good condition, agency warnings were taken less seriously, and the media approach diffused public pressure for a military "rescue" operation.

Many governments had an interest in discounting the existence of "missing" refugees. Having rejected calls at the start of the crisis for military intervention to disarm militias in the refugee camps – a move that might have led to a far earlier return – some were happy to abandon plans for a military operation which faced enormous logistical challenges.

With memories of past, problematic interventions – especially Somalia – still fresh, they also feared domestic political fallout from going in, especially if the operation was a failure. The Rwanda government would hardly have welcomed a multinational force, since it would have affected the classic balance of power politics in the region.

Yet reports of hundreds of thousands of people heading west continued to trickle through from eye-witnesses in Zaire, until their existence was eventually accepted. It became clear that they were not only very real but also in desperate need. In early 1997, Sadako Ogata, the UN High Commissioner for Refugees, visited several sites where refugees had temporarily settled, as did Emma Bonino, European Commissioner for humanitarian affairs, who said she had "found the refugees that didn't exist" and estimated that 200,000 to 400,000 remained.

As well as being a potent warning about the hazards of numbers, especially if agencies do not speak with one voice, the episode marked a significant shift in media/aid-agency relationships, demonstrated the new political atmosphere against intervention – especially in Africa – and may have opened a new era in which agencies cannot assume that their statements will be trusted by public or governments, let alone acted on.

---

that used by the UN World Food Programme and the UN High Commission for Refugees (UNHCR)for protracted refugee operations, would measure more accurately the costs of responding to long-term crises. It would also facilitate sustained monitoring of international crisis financing. Stripped of the large and fast fluctuations during the acute phases of high-profile emergencies, the size of the real "humanitarian deficit" would become much clearer.

Another reason to demand better figures is that some donors are categorizing at least part of their spending on post-conflict reconstruction – such as in the former Yugoslavia – as "emergency" aid or "relief". This type of reporting further conceals the fact that many populations in need are receiving less relief aid, not more.

The financial picture may be unclear, but the state of the humanitarian system has been examined in important evaluations of major relief operations in the Great Lakes, Liberia and Sudan. These evaluations, other studies and the extensive operational experience of many individuals and agencies are helping define the agenda confronting the humanitarian community on the eve of the 21st century.

Limits of action

"Relief cannot substitute for political action" is the primary conclusion of the Rwanda evaluation. This striking, and seemingly obvious, conclusion is of profound importance based as it is on extensive analysis of the political and humanitarian response to genocide and its aftermath. It implies that the boundaries of humanitarianism must be drawn carefully: Much of what relief agencies are blamed for is in the domain of governments and other political bodies.

In all evaluated emergencies, political violence and its effects, not the quality of relief programming, are primarily responsible for causing human misery. With important exceptions, each evaluation concluded that relief had made an important contribution towards saving lives, and had helped protect conflict-affected populations. These findings seem to be generating remarkably little change at the political level. Instead, as in the case of eastern Zaire, media attention remains focused on the perceived or real deficiencies of relief operations, not of political and military response.

Dynamics of war

Disillusion with humanitarian relief among politicians, the media and perhaps the public in the North frequently reflects another finding of recent studies. Control over apparently benign goods such as food aid, potable water and drugs is necessarily controversial in zones of armed conflict. Different groups win or lose, depending on how relief resources are distributed. Warring parties make neutral and impartial relief programming, where only the victims benefit from aid, extremely difficult in armed conflicts. In many conflicts, civil society is politicized and parts of it militarized. In Rwanda, for example, the genocide was perpetrated by civilians, including women, as well as by military and paramilitary personnel. Such situations raise very difficult questions regarding relief agencies' choice of partners in conflict zones and the way in which they negotiate with belligerent parties to secure access.

From Goma to Khartoum and Monrovia, research has highlighted the risks of both failing to create an appropriate political framework for the delivery of humanitarian supplies and the poor management of such supplies.

Drawing on this analysis, an unpalatable conclusion is being drawn by some critics, such as Alex de Waal of the UK-based African Rights group. Rather than working

*Figure 5.2
Annual levels of
emergency aid
(excluding food
aid).*

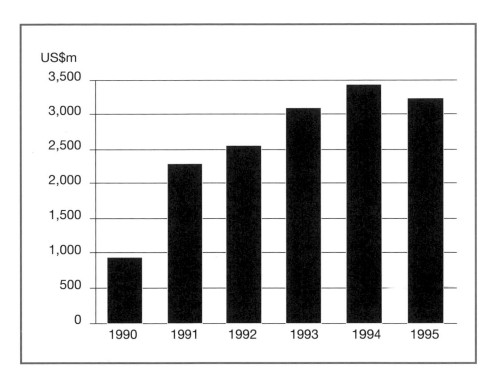

harder to maintain the integrity of relief in wartime, they ask if perhaps we should not provide relief at all. Such an approach risks reinforcing the isolationist tendency of the major powers, and justifying both political and humanitarian disengagement from conflict-related crises. Such an analysis represents an attack on the fundamental principles of humanitarianism and concern for human rights.

Others, such as Mary Anderson of the Collaborative for Development Action, in discussing how agencies should first "do no harm", and John Prendergast advocating "aid with integrity", have built on the work of those such as David Keen and Ken Wilson to urge that as relief may feed into conflict, perhaps it can also be used to provide for conflict-affected populations. These analysts are suggesting options for agency action that are culturally sensitive, politically sophisticated and offer new ways of working within the constraints of falling funds and rising complexity.

Contracts and rules

Countering the disengagement arguments are those who seek to defend the humanitarian imperative, while acknowledging the failures and limitations of the current humanitarian system. Considerable discussion and operational innovation has centred around mechanisms for enhancing regulation of humanitarian assistance, and in some cases regulating the conduct of war. In addition to the promotion of the *Code of Conduct for the International Red Cross and Red Crescent Movement and NGOS in Disaster Relief* (see Chapter 11), in countries such as Sudan and Liberia there have been important attempts to develop contracts with belligerent parties to urge respect of humanitarian law and so increase the quality and quantity of humanitarian space.

However, these debates bring a distinct sense of *déjà vu*. For over a century, the International Red Cross and Red Crescent Movement has worked to defend the Geneva Conventions and sought to link relief with an attempt to prevent the worst excesses of war. Reinventing (or at least promoting more widely) this particular process is vital to maintain the integrity of the humanitarian system against accusations of poorly-managed relief programmes.

Two challenges stand out. First, what sanction is the international community willing to use in the face of widespread violations of international law? Second, what mechanisms are in place to monitor adherence to standards of humanitarian programming? Donors have been slow to make clear their expectations of NGO adherence to humanitarian standards. Concern for financial reporting still seems to predominate over concern for principles of neutrality and impartiality.

Reforming the system

The international humanitarian system needs reform and leadership. The UN is subject to increasing scrutiny about the quality and relevance of its humanitarian strategy in the post-Cold War era. The mandates of its specialist agencies look tired. For example, more people are displaced internally by war than become refugees, and there are many other causes of displacement, yet there is still no policy or single focal point to ensure they are protected. Many UN agencies claim a role in other areas, such as food security and health, causing duplication of effort and blurred responsibilities when things go wrong. Creation of the UN Department of Humanitarian Affairs (UNDHA) in 1991 has not resolved the lack of coordination between the different specialized agencies, what some in the UN call its hollow core. Nor does UNDHA seem to have adequately met the requirement of providing a bridge between the humanitarian and political spheres, though this may be less due to UNDHA than agencies and governments proving resistant to being coordinated. These failures have led some to urge a merger of the relief functions of UN specialized agencies and the creation of a Security Council humanitarian subcommittee. A new UN Secretary-General has added impetus to the calls for reform, and these issues are due to be discussed in 1997 at the UN Economic and Social Council.

> *Donors have been slow to make clear their expectations of NGO adherence to humanitarian standards.*

Retrenchment or reform?

Any reform of the humanitarian system will be determined by politics, which will also affect how much money is made available to address the effects of conflict, refugee flows and other crises. If 1996 has been a good year for analysis of the problems confronting the humanitarian community, the rigorous self-criticism should be clearly and carefully communicated to mount a vigorous defence of humanitarian values. Such a defence will have at least two components.

First, making sure that the standards of humanitarian programming are improved, and in particular that care is taken to maximize its impact and effectiveness. In large part, this will require ensuring the primacy of humanitarian values of neutrality and impartiality in relief programming.

Second, the humanitarian system needs a coherent advocacy agenda which sets out clearly what it can and cannot achieve. Relief cannot resolve problems of underdevelopment or the causes of conflict, still less provide a substitute for a judicial process. What it does and can do is save lives and help reinforce people's dignity. The humanitarian community cannot do the work of others, but it can advocate for the work to be done.

Chapter 5 *Sources, references and further information*

Anderson, Mary. *Do No Harm. Supporting Local Capacities for Peace through Aid.* Cambridge, Massachusetts: The Collaborative for Development Action, 1996.

Apthorpe, R. et al. *Protracted Emergency Humanitarian Relief Food Aid: Towards "Productive Relief". Programme Policy Evaluation of the 1990-1995 period of WFP-assisted refugee and displaced persons' operations in Liberia, Sierra Leone, Guinea and Côte d'Ivoire.* Mission Report, Rome, 1996.

de Waal, Alex. *Sorry St Bb: It's time we banned aid.* The Observer, London, 1996.

DANIDA/ODI. *Joint Evaluation of Emergency Assistance to Rwanda.* Five volumes, London and Copenhagen, 1996.

Karim, et al. *Operation Lifeline Sudan: A Review. UN Department of Humanitarian Affairs.* Geneva, 1996.

Keen, D. and Wilson, K. *Engaging with Violence: A Reassessment of Relief in Wartime.* In Macrae, J. and Zwi, A. (eds.): *War and Hunger: Rethinking International Responses to Complex Emergencies,* pp.209-221. London: Zed Books, 1994.

Macrae, Joanna. *The origins of unease: setting the context for current ethical debates.* Background paper prepared for the workshop on the ethics of humanitarian aid, ECHO/VOICE, Dublin, December, 1996.

OECD. *Development Cooperation: Efforts and Policies of the Members of the Development Assistance Committee.* Paris, 1996.

Richards, Paul. *Fighting for the Rainforest; War, Youth & Resources in Sierra Leone.* London: James Currey, 1996.

Richards, Paul. *Small Wars and Smart Relief; Radio and Local Conciliation in Sierra Leone.* In the Creative Radio for Development Conference Report, Health Unlimited, London 1996.

Web sites

ICVA: http://www.icva.ch/

InterAction: http://www.interaction.org/

OECD (Washington): http://www.oecdwash.org/

UN: http://www.un.org/

UK ODA: http://www.oneworld.org/oda/

USAID: http://www.info.usaid.gov/

Tackling hunger, disease, and poverty in Somalia requires the medicine of peace and security, not just aid. Local efforts in service provision are impressive, but where is the coherence of action that turns those efforts into national change?

Somalia, 1994.

Chapter

6 Somalia: aid where there is no government

The challenge of Somalia is clearly not one of a lack of central government, but one of a lack of peace and security; today it has a plethora of local authorities, religious courts, governors, self-proclaimed republics, councils of elders and more, with little coherent national or regional coordination, and none is able to secure and then guarantee a permanent end to inter-clan violence.

The country also has hundreds of thousands of refugees and displaced people, deep poverty, massive inequalities, poor health standards, limited education, and a situation that offers young people growing up few options and plenty of guns. Providing the drought and threatened famine can be managed well in 1997, past crises are long gone, yet the usual next step of rehabilitation seems hard to take amid such insecurity.

Facing these realities, for humanitarian agencies and their donors the question must be a simple one: How can they help ensure that the needs of the most vulnerable are met in both the short- and long-term amid the persistent violence that accompanied the sudden return to a stateless society of diverse power centres and shifting politics.

For return it has been, since Somalia experienced just nine years of fragile Western-style democracy, back in the 1960s, between more than 70 years of colonialism and just over 20 of dictatorship that ended when Mohamed Siad Barre fled in 1991. Despite repression, plenty of war and sudden political lurches during the Cold War, nothing in the last century or more has fundamentally undermined the one certainty of Somalia: the clan structure of a nomadic warrior society based on livestock herding in an unforgiving arid land.

> *It made perfect sense that the warlords would fight over relief food.*

In traditional Somali politics, no leadership is fixed; all decision-making is conducted by segmented groups of fiercely-individualistic kinsmen whose patterns of alliance and confrontation are fluid. Pastoralists meet in assemblies where all adult male family heads – the elders – seek consensus rather than voting. In this deliberately and defiantly self-reliant, uncentralized society, there is no tradition of a unified state, no set political offices or ranked leaders. Economically, in Somalia's history, there was also no basis for central state authority.

Somalia today may be condemned as a "failed state", but that implies its predecessor was some sort of success. From independence in 1960, centrally-provided services in the country were funded almost entirely by superpower-driven foreign aid, leading to a structure that collapsed not long after the end of the Cold War. Apart from a Latin written script for the once solely-oral language, the regime left behind only well-armed opposition groups, ready to kill each other and anyone else who got in the way as they fought for territory and loot.

The violent complexities of Somalia continue to disturb Western governments, including the donor community, which first found the stateless country an unpleasant concept, and recoiled further at its reality when all external attempts to reconstruct central authorities were ignominiously torn down. Since the Americans pulled out of Mogadishu and the United Nations (UN) left in 1995, most of the world – including the UN Security Council – has all but ignored Somalia.

US author and former aid worker, Michael Maren, found confusion in Somalia: "Neither the Americans nor the UN ever seemed to get a firm grasp of the clan system. The walls of their offices were plastered with clan diagrams torn from academic books...the diagrams broke the system down into clan families, subclans, sub-sub-clans. It was all very neat and very graphic. But the Westerners tended to look at the family trees as if they were corporate organizational charts. They therefore concluded that power emanated from the top, and that everyone was really part of one big family. The fighting was internecine and therefore senseless.

"What they never seemed to understand was that the Somalis themselves never thought in terms of organizational charts. Their perspective on their own lives was from the bottom up. Starting with immediate family and climbing up the family tree, the farther away you got, the more remote the connection. The loyalty of the foot soldiers to the chief at the top of the clan lasted only as long as the spoils of war came down through the ranks.

Box 6.1 Somaliland: `republic' without recognition

At a distance, it can be easy to forget that "Somalia" is at least two countries. The pre-independence split between the north-western section, that was a British protectorate from 1886 until 1960, and the rest, that was Italian, has returned in the form of the territory that has even readopted the name Somaliland.

A declaration of independence in 1991 does not stop disputes about how much of the territory the new administration can claim as accepting any form of national authority. Nor does it stop disagreements about whether Somaliland should be independent or in a federated state, and it cannot prevent its one million people suffering occasional upsurges in clan-related violence.

Although Somaliland was an early starting point for much of the resistance to Siad Barre, which made it a target of military action and political repression well before the war of the 1990s, it has pulled together, from a very low base, a range of national and local structures, held a national congress about its future, and made efforts in governance, security and development.

Unfortunately for the people of Somaliland, creating some sort of state, a measure of peace and a little prosperity, has brought absolutely no international recognition, very little aid except through NGOs, and almost no attention, as most eyes – media, donors, diplomats, aid staff – watch Mogadishu.

Recognized or not, the self-declared "republic" of Somaliland has a police force, army, courts, taxes, a range of busy, local NGOs, and plenty of even busier businesses, many of which have links with the continuing economic success story of Berbera port and its massive livestock exports.

Berbera benefits from the Cold War heritage of Somalia, with both Russian- and American-built warehouses and docks. Despite problems of limited water and electricity supplies – which the European Community is tackling – and the need to improve veterinary checks and vaccination to avoid animals being returned, Berbera is booming. Its 1995 (and partial 1996) figures show exports to the Gulf states and beyond of 22,000 (38,000) camels, 75,000 (59,000) cattle, 2,684,000 (2,134,000) sheep or goats and 461,000 (170,000) hides. Part-year 1996 figures show imports of: sugar 62,900 tonnes; rice 95,400 tonnes; wheat 21,500 tonnes; cooking oil 4,400 tonnes; diesel oil 6,200 tonnes; and 3,345 vehicles.

This success means hard currency for the government, jobs for port workers, opportunities for truckers and herders, and profits for traders. The animals come from all over the region, including Ethiopia and Somalia, grazing as they go, and making their way through hundreds of miles that many outsiders would assume were still at war.

Providing it is not overtaxed by an administration with few income sources, Berbera Port Authority has good prospects. Its nearest rival, Bosasso, in the almost equally-peaceful Bari region in the north-east, takes smaller ships and faces silting problems, while the only other Somali deep-water port, in Mogadishu, has been battled over by factions.

But a revived livestock sector, low-tax trade and the prosperity of some is no guarantee that vulnerability is really being reduced. Indeed, aid observers see many of the same elites from the 1960s now in positions of power, a widening gap between rich and poor, especially in rural areas, and the economy dominated by half a dozen wealthy businessmen.

The president and ministers in the capital Hargeisa worry about independence and money, which leads to problems with aid agencies, the only entities from donor nations: threats of expulsion, demands for international recognition, and accusations of double standards, given the founding of new states in Europe.

Despite these difficulties – not present in government-free Somalia – the comparative peace and stability of Somaliland mean that children can be immunized and go to school, health services can be rehabilitated, water supplies improved, land-mines cleared, and malnutrition and hunger tackled.

Meanwhile, there is the instability of a very soft new national currency – more problems for the poor, with little access to dollars or the hard Somali shilling – and long-term issues of land tenure and privatization in a country facing the usual conflicts between nomadic pastoralists and settled farmers.

"The battles fought in Mogadishu were for scarce resources, real wealth. The stakes were high. For the last ten years of the Barre government, relief food poured through the docks of Mogadishu. The government used that food to build its powerbase and control people in the hinterlands. Individuals used the food to get rich. It made perfect sense that the warlords would fight over relief food. Food was power and survival for the family.

"It was the petty nature of the Somalia dispute – small men lusting after power and loot – that made the conflict so intractable. There were no issues. No ideological differences. Nothing to negotiate. While the UN was exerting its energy to broker an agreement between the warlords, [the factional leaders] were worried about keeping their relatives happy. In other words, billions of dollars and several hundred-thousand foreign nationals were involved in a global operation to settle what was at its core the politics of dysfunctional families."

However, once outsiders gave up trying to tell Somalis what to do, and stopped fuelling the fighting through massive cash injections into clan-linked operations and the legitimizing impact of anti-faction military action, a new impetus was given to real and potentially-sustainable efforts to put back together not a state, but the local and sectoral structures to deliver what people need. Using traditional peace mechanisms (discussed in detail in the *World Disasters Report 1994*), local and sometimes regional agreements (see Box 6.1) are being reached in many parts of the country to allow at least a semblance of normal Somali life, especially trade, farming and livestock herding.

> *It was the petty nature of the Somalia dispute that made the conflict so intractable.*

Clan struggles continue, and cannot be ignored for either their occasional violence or the endless manoeuvrings by elites – political, military, religious, business – over the minutia of alliances that decide the Somali political and security environment, and may, some day, result in a longer-term resolution of the crisis.

Factions often bypassed local governance for the illusion of national status and the reality of controlling an exploitable asset, be it a port, airfield or crossroads. This has allowed others to pursue issues that matter to people without guns: law and order, systems to settle disputes and regulate markets, finding the money for schools. Local communities have devised their own structures for governance that suit their specific situation. In many cases, notably in north Mogadishu, Islamic courts dispense Shariah law, with its potentially-harsh – to Western sensibilities – penalties, such as hand amputation for theft, but exceptions for those in need.

In villages, in urban neighbourhoods and on the pastoral range, these grassroots political structures are diverse in scope and effectiveness, and fluid in nature, but they meet the local needs of communities, while a centralized structure – or even systems for broader regional or national coordination on issues of health, for example – looks like taking a long time to get established.

Land struggles

Although front-line aid workers need to understand enough of the politics to deal with their practical consequences, humanitarians are in Somalia to work around the gunfire and politics, not to create governments, choose sides or pass judgement on political structures.

Aid agencies clearly do need to know who is winning and losing in conflicts, in both the short and long term. Discussing the military intervention, US analysts Walter

Clark and Jeffrey Herbst observed: "The intervening forces failed to recognize which Somalis had been victims. Much of what appears to be incomprehensible warfare in Somalia is a struggle for land between the African farmers in the south and the northern, clan-based nomadic groups, which are better armed.

"Most of the victims in Somalia were members of the Bantu and Benadir clans, sedentary coastal groups who traditionally live in uneasy coexistence with neighbouring ethnic Somali groups, and the Rahanweyn clan, who work the rich agricultural land in the Juba and Shabelle river valleys in the south and which is the weakest militarily."

That helps puts into context, for example, recent claims that some forces have moved into that rich agricultural land to tax the 30,000-tonnes-a-year banana trade – the second largest export of Somalia – to fund arms purchases, and battles involving the Rahanweyn Resistance Army over what it sees as the factional occupation of the southern city of Baidoa.

Box 6.2 **Famine and flood, factions and forecasts**

Drought and famine are back on the agenda for Somalia and much of East Africa in 1997. The UN Food and Agriculture Organization warned that the secondary "Der" crop in Somalia, which accounts for a quarter of annual cereal production, was 60 per cent below 1995/96 and that donors should make contingency plans to send food.

Faction leaders appealed for international help, only to have the Somali Aid Co-ordination Body (SACB) of foreign donors and aid agencies declare sharply that faction leaders themselves will be responsible if people starve: "The onus for ensuring freedom of access to affected areas, guaranteeing the security of aid personnel and allowing the free movement of goods rests wholly with the Somali people and their leaders.

"The international aid community can only react to the current drought conditions, if the necessary measures are taken by the Somali population. Primary responsibility to respond to the distress of the most vulnerable members of the Somali population rests with the Somali leaders. An appropriate response cannot come solely from the international community."

Although overall cereal production for both seasons in the 1996/97 agricultural year was 10 per cent higher than in 1995/96, the cumulative effect of two years of mediocre harvests pushed up food insecurity in southern Somalia. The most vulnerable populations are rain-dependent farmers, herders-already affected by high cattle mortality, and urban populations in southern Somalia who have found their purchasing power eroded by seasonally-high cereal prices and depreciation of the Somali shilling.

Worried that a failure of the 1997 Gu rains, due to start in April, could prompt a major food security problem, especially in southern Somalia, agencies were developing contingency plans for the purchase of cereals from Ethiopia, positioning food stocks in secure sites in Somalia, and closely monitoring market conditions.

Ironically, the drought coincided with floods in the Juba valley, which affected 57,000 people, sweeping away crops and food stores. Caused by high rains in Ethiopia, the floods were worsened by years of poor dyke maintenance, and drought-hit farmers breaking river banks to irrigate their fields. Local people are working with ICRC, UNICEF and other agencies on flood-control measures.

Whatever the reason for their hunger, many poor Somalis will rely on traditional coping mechanisms to survive: eating their animals and wild fruits, roots and leaves, moving to search for work, and trading in livestock, firewood, honey and other products. Many families will also rely on remittances from members working abroad or elsewhere in the country.

Based on the harvests, farmer stocks, and coping mechanisms, the US Agency for International Development's Famine Early Warning System (FEWS) estimated the national cereal gap in January-July 1997 at 58,436 tonnes. The regions of Bay, Gedo, Lower Shabelle, Bakool, and Lower Juba will be most affected. FEWS recommended that food aid be sold at market prices through local auctions and that food-for-work only be used in more severe areas.

For aid agencies there may also be resonances when Clark and Herbst discuss reconciliation and negotiating with armed groups: "The challenge is particularly difficult because promoting long-term reconciliation may mean empowering the unarmed, while short-term exigencies will require reaching a *modus operandi* with the warring factions."

Somalia has been a watershed for donors and agencies, forced to rethink their operational methods, mandates and structures. But not all questions and doubts have been resolved: where and with whom to work; how to work in ways that reduce the risks for beneficiaries, local staff and expatriates, while not abandoning those in greatest need; guarding against the politicized environment undermining principles like impartiality and neutrality; the values and standards needed to guide those working where there is no government.

One step back from the front line, donors are also being urged to come to a clear understanding of local political structures and determine how much aid can be channelled through them, while addressing concerns that aid which strengthens local authorities risks promoting clans and reducing the chances of a return to centralized authority, or just greater coordination.

The aid environment for Somalia is very different from the 1970s and 1980s, when bilateral assistance served obvious foreign-policy objectives. The global total available is falling, and has to be stretched across many more countries and crises, including parts of Europe and the former Soviet Union, inaccessible a decade ago. Donors are looking for better results, adding more conditions and expecting far more self-reliance from beneficiaries, though they are now willing to put funds through both international and local non-governmental organizations (NGOs).

> *Somalia has been a watershed for donors and agencies.*

Donors are committed to rehabilitation, but insist that violence can no longer be tolerated. The European Union (EU), announcing in 1996 a two-year, 60 million US dollar rehabilitation and development programme, said money would only go to regions "where peace and security prevail, and where there are responsible Somali authorities" to administer the aid. Funding covers productive activities (livestock, agriculture and fisheries), social sectors (health, water, sanitation and education), revenue generation, job creation and training. Even without a state, the EU clearly thinks there are Somalis with whom it can do business.

Most UN specialized agencies are also developing ways to work in Somalia. Despite a poor response in 1995, the Consolidated InterAgency Appeal in late 1996 called for more than $100 million for work up to December 1997 in emergency response, reintegration of displaced people and returning refugees, rehabilitation, and governance. It set out the need for a differentiated strategy to cope with the problems and opportunities of each local area.

In all there are about 40 UN agencies and NGOs working in Somalia in health, education, agriculture and other sectors, linked through a range of bodies, such as health coordination groups, which include Somali health professionals. The UN has established a data bank on Somalia, while the World Food Programme and the European Commission set up a Food Security Assessment Unit to monitor harvest, food prices and nutritional data.

In their health development programmes in Somalia, the UN and international aid agencies, most of which are sponsored by the EU, are involved in many programmes, such as rehabilitation and development of orthopaedic centres, training physiotherapists and orthopaedic technicians, production of orthopaedic equipment and starting wheelchair production units. Regional medical stores have been established to facilitate pharmaceutical distribution, and cost recovery for drugs has been introduced with a view to eventually commercializing drug procurement and distribution.

> *Somalis are creating new organizations and networks wherever peace and security provide any space.*

The largest development agency, the World Bank, finds the country harder to cope with. It cannot itself implement projects to rebuild war-torn societies, but where governments exist, it aims to "promote a secure and stable environment for development" through such programmes as demining, demobilization of fighters, and the reconstruction of infrastructure. But the World Bank is a bank, and it wants the prospect, if not the reality, of political stability to ensure its loans will be repaid, and "properly-constituted authorities" to sign its contracts.

Not that some Somalis are short of money, with the economy kick-started by looting and the cash inflow from the American and UN's military and aid operations. No government means no bureaucracy and limited taxation for businesses and local NGOs. Trade in Somalia has flourished as businessmen, who are usually also bankers and brokers, buy and sell across internal and international boundaries with their radios and satellite telephones. Exports of livestock and fruit cross over with imports of medicines, weapons and the narcotic plant, qat, chewed by many Somalis.

With no new notes being produced and backed by nothing more than people's confidence, the Somali shilling has retained its value alongside the ubiquitous US dollar. Somalis working abroad are sending back enough money, it is estimated, to finance as much as 60 per cent of urban consumption through individual families.

Much is working in Somalia, including ports and airports large and small, while the new realism of aid agencies and donors, particularly the EU, has brought efforts to work with the economy, such as by helping vets – once a government-subsidized service – to set up sustainable businesses charging the full costs of drugs and advice.

Even in Mogadishu, there are plenty of hard-working companies, from pasta manufacturers to computer trainers, and including the Al Furqaan boatbuilders who kept on working through the troubles, the Barakaat telephone company, which employs 350 people, and the generator entrepreneurs who offer street lighting as a bonus for customers.

And beyond the business world, Somalis are creating new organizations and networks wherever peace and security provide any space, such as the coordinating Mogadishu Joint Health Authority, and doing their best to maintain or revive older institutions. There is still, for example, an active Somali Olympic Committee, which maintains that sport, as a unifying force, could help peace building.

Red Crescent respect

The International Committee of the Red Cross (ICRC) has been active in Somalia since 1988, when its main focus was on the north-west, where it ran a surgical hospital in Berbera and set up a large relief operation for Ethiopian refugees. The International Federation of Red Cross and Red Crescent Societies has had a presence since 1990, supporting, in a long collaboration with the Norwegian Red Cross, the activities of the Somali Red Crescent Society (SRCS).

In 1991-92 the International Federation withdrew because of the conflict, but in September 1992, it was decided that both the ICRC and International Federation would work in Somalia under their own different mandates, an arrangement still in force today, reflecting the mix of peace and conflict in much of the country.

The International Federation aimed to support the SRCS in a health programme for women and children, and to strengthen the operational capacity of the National Society in less-conflictual areas. ICRC operates in Mogadishu and other areas still affected by conflict. The International Federation operates in areas which are considered more stable.

Up to the end of 1993, the ICRC carried out one of its biggest-ever relief operations, at one time feeding more than a million people a day. Now, the emphasis is on health care via hospitals and health posts in areas of conflict or displaced people, and more development-oriented programmes at local level, including fisheries, agriculture, veterinary assistance, and the rehabilitation of wells and river dykes. Its seed programme reached 200,000 people in 1996, and efforts to combat the threat to livestock of tsetse fly have begun.

Somali health standards are among the worst in Africa. In 1992 average life expectancy at birth was 46 years and the infant mortality rate was about 123 per 1,000 live births. The SRCS, in the absence of a functioning Ministry of Health, has had to step into the breach, and finds itself – despite the closure of some centres from violence – running the country's only indigenous health service. The 24 clinics in Garoe, Kismayo, Baidoa and Mogadishu are supported by the International Federation and donor National Societies.

This network, far more than the system the previous government operated, serves a population of more than 350,000, and records an average of 30,000 visits a month. In addition, the SRCS, with the support of ICRC and the International Federation, runs hospitals in north and south Mogadishu and, with the Italian Red Cross, the only referral hospital in the north-east, at Garoe. Efforts are being made in Garoe to transform it into a community-based hospital, with the local authorities appointing a committee of elders and local officials.

> *Somali health standards are among the worst in Africa.*

Such an important health network could not successfully function without the SRCS, which is able to work all over the country. Where expatriates are unable to travel to health centres for security reasons, Somali Red Crescent staff are able to monitor and support the programme. The presence of expatriates generally creates high expectations, but SRCS staff can have a lower profile. Where problems arise in Somalia, the national staff of the SRCS are alert to the possible reactions.

As the only indigenous institution still functioning on a national basis, the Somali Red Crescent Society has earned considerable confidence and respect both among the various political factions and ordinary Somalis. Throughout the conflict, the SRCS played a crucial role in making possible the distribution of food and other supplies to needy groups. Its volunteers and staff continue to involve the community in its work. Health education is a major activity, with training for midwives and traditional birth attendants, first aid for school students and work on HIV/AIDS awareness and other issues.

Junk weapons

Somalia is a good example of a place full of what British military historian John Keegan has called, in an allusion to fast food, "junk weapons". Traded around the world in large numbers and easily obtained by any violent group, these small arms are cheap and plentiful, yet their capacity to disrupt entire societies as they maim and kill is enormous.

In Mogadishu, continuing fighting among factions for outright control of the city has seriously affected health programmes, and rehabilitation and development projects. Extortion is common in a city still awash with weapons and high unemployment. In the last few years, more than a dozen foreign aid workers have been kidnapped and held to ransom.

Despite the respect in Somalia that allows it to work in even the most unstable areas, the SRCS has not been immune to the violence. During the war, all its property was looted and destroyed. In 1996, indiscriminate shelling in Mogadishu killed three of its workers and wounded more than 20. The other components of the International Red Cross and Red Crescent Movement have also fallen victim: the ICRC and the International Federation now run their Somali delegations from Nairobi.

> *Violence is now an expected part of the disaster-response environment.*

The violence in Somalia forced agencies to hire armed guards in order to carry out their work in safety. It was a difficult and unprecedented step for the International Federation and ICRC, with implications for the declared neutrality and impartiality of all parts of the Movement.

Although the practice of only travelling with armed guards has virtually ceased, with relative security in most parts of the country, apart from Mogadishu and Baidoa, most agencies still hire guards to protect premises.

Given the obligation to assist the most vulnerable, a guard employed by the ICRC or International Federation protects supplies and personnel. But the guard is first and foremost a clan member, and few are impartial, so his mere presence may provoke an attack from members of another clan, who cannot conceive of an organization that is neutral and impartial, and see the guard as an expression of bias.

The reality and perception of neutrality is not easy in a country with so many different allegiances. Recruitment involves balancing clans in jobs so organizations are not accused of bias. In long-term work, all clans have to be included, making it extremely difficult to balance programmes to assist the most vulnerable. Single programmes focused on one area are almost impossible to sustain.

These factors put tremendous constraints on agencies trying to operate in Somalia, even when projects are sited in areas of comparative calm, and on the types of projects that can be implemented.

Conflict options

Somalia reflects the general growth of violence – from criminal attack to terrorism and all out war – which is now an expected part of the disaster-response environment of aid agencies. The violence of Somalia may be more extreme or frequent, but it is hardly unusual, since many aid agencies intervene in situations they might well have avoided only ten years ago, from Bosnia to Liberia.

Apart from the issue of staff security, violence is a swift method of producing winners and losers, and the losers are usually of humanitarian concern. While many NGOs pursue ideas of conflict management or resolution, there is a case that agencies also need to look for the kinds of strategies, tactics and tools that can work even in violent environments, because they are simple, or highly mobile, or "hardened".

Despite constant risks, aid agencies have continued to provide support to the most vulnerable segments of the population in Somalia, wherever conditions of access and security allow. In most Somali regions, rehabilitation interventions have had to favour the smaller ad hoc projects rather than district or regional reconstruction and development programmes, due to continued insecurity and lack of progress in political reconciliation process.

Agencies' Somali experience suggests a range of factors that could help projects and programmes to have a better chance of success in meeting the challenge of Somalia and other failed states. These factors include:

■ Prospects will be improved by having local partners to help design and manage the work, and direct contact with beneficiaries or projects that are effectively beneficiary-run, through factors such as skill transfer and the training of trainers. Final ownership is very important.

■ Unless the location is fully secure, or the work is conflict resistant, the project will probably benefit from being flexible, low cost, low tech and small scale; and it would be wise to build in a factor for damage or losses.

■ Projects that have been tested and adapted to local circumstances obviously have a greater chance for survival, while those that meet clear, community-defined needs, and equitably share the benefits, should draw a broader support, though that may mean less targeting of resources.

■ The project should be allowed to develop at a pace that suits those who have to run it, rather than the external demands of donor finance, agency review, media coverage or evaluation.

■ Even, perhaps especially, in a failed state, there are many other "actors", from private companies to local farmers, craftspeople to religious groups, who should be considered or consulted to ensure the project is wanted, workable, useful and sustainable.

■ Unless confident of security, agencies should not fuel conflict by bringing in resources – or expatriate staff – if these are likely to encourage factions to fight over them, or if they could add new inequalities to those that helped cause the initial conflict.

■ And, in reverse, explore opportunities for more equitable means of distributing resources, such as radio broadcasting of education and information materials.

These suggestions also expose a dilemma in working within failed states. More than in settled situations, these areas of conflict, or places that lack legitimate authority, demand very close attention to project design and implementation, yet for sustainable success they almost certainly need less external participation and management.

Somalis may not want a central government and its trappings to return, but local initiatives and efforts clearly have a limited impact if there are no structures or systems for regional or national coordination, while the costs and complexity of assistance rise. Some people are doing well amid insecurity, many are not. More fighting means less jobs, less farming, and far more hunger, poverty and disease.

But failed states do not mean passive people, and Somalis are unlikely to wait for outsiders to solve their problems, if they can see another way. In 1996, 1,000 displaced people living in Kismayo, and keen to return to their farmlands ahead of the rains, gave up on stalled transport arrangements. Rather than wait in frustration, they picked up their possessions, set off and walked for a week to get home.

Chapter 6 *Sources, references and further information*

Clark, Walter and Herbst, Jeffrey. *Somalia and the Future of Humanitarian Intervention.* Foreign Affairs, March/April 1996.

Crocker, Chester. *The Lessons of Somalia; Not Everything Went Wrong.* Foreign Affairs, May/June 1996.

Doornbos, Martin et al (eds.). *Beyond conflict in the Horn: the prospects for peace, recovery & development in Ethiopia, Somalia, Eritrea & Sudan.* Lawrenceville, NJ: Red Sea Press, 1992.

Hirsch, John and Oakley, Robert. *Somalia and Operation Restore Hope: reflections or peacemaking and peacekeeping.* Washington: United States Institute of Peace Press, 1995.

International Cooperation for Development. *Building Partnerships for Participatory Development; Report of a Workshop held in Hargeisa, Somaliland.* London, 1996.

International Federation Somalia Delegation. *Annual Report 1996.* Nairobi, March 1997.

Kusow, Abid Mohamed. *Somalia's Silent Sufferers.* Africa News Service, January, 1993.

Lewis, I.M. *Understanding Somalia; Guide to Culture, History and Social Institutions.* London: Haan Associates, 1993.

Lyons. *Somalia: state collapse, multilateral intervention, & strategies for political reconstruction.* Washington: Brookings Institution, 1995.

Maren, Michael. *Somalia; Whose Failure?* Current History, January, 1996.

Maren, Michael. *The Road to Hell; The Ravaging Effects of Foreign Aid and International Charity.* Free Press, 1996.

Minear, Larry et al. *The News Media, Civil War & Humanitarian Action.* Boulder, Colorado: Lynne Rienner Publishers, 1996.

Omaar, Rakiya and de Waal, Alex. *Saving Somalia Without the Somalis.* Africa News Service, January, 1993.

Physicians for Human Rights and Africa Watch. *Somalia: no mercy in Mogadishu: the human cost of the conflict & the struggle for relief.* Boston: Physicians for Human Rights, 1992.

Sahnoun, Mohamed. *Somalia: The Missed Opportunities.* Washington: United States Institute of Peace Press, 1994.

Somalia – Still Breathing, in its Way. *The Economist.* London, August 31, 1996.

Somalia Country Report. *The ACP-EU Courier.* Commission of the European Communities, Brussels, March/April 1997.

Web sites

Africa Online: http://www.africaonline.com/

Africa News: http://www.africanews.org/africa.html

Foreign Affairs: http://foreignaffairs.org/

ICRC: http://www.icrc.org

International Federation: http://www.ifrc.org

NomadNet: http://www.interport.net/~mmaren

Somalia materials: http://ralph.gmu.edu/cfpa/peace/clarke/

Somalia links: http://strix.udac.uu.se/insts/antro/bh/Somalia.html

UNDHA ReliefWeb: http://www.reliefweb.int

USAID Famine Early Warning System: http://www.info.usaid.gov/fews/fews.html

The hazards of the Caribbean are real and multiple, disaster preparedness is possible but efforts are still needed to bring together the work of researchers, planners and aid agencies to make more of the opportunities to save lives and livelihoods.
Montserrat, 1996.

Chapter

7 High-cost hazards of the Caribbean

The Caribbean has weathered the two worst consecutive years on record of hurricane activity with early indications of an above-average 1997 hurricane season. The "Ring of Fire" of some 30 Eastern Caribbean volcanic sites continues to be violently active on Montserrat. The region faces many other hazards. Hardly surprising that the 1997 Caribbean Insurance Conference included a discussion ironically entitled "The Caribbean, A Disaster Area – Let's Evacuate".

Evacuation is not a real option for most. Scientists at the Seismic Research Unit (SRU) on Trinidad think that the tiny Dutch territory of Saba in the Leeward Islands is probably the only island in the region where a volcanic eruption or other seismic event might force everyone to leave.

The vulnerability of the region to a wide range of natural and technical hazards – from high winds, floods, volcanoes and earthquakes to oil and chemical spills – emphasizes the need to improve early warning, community preparedness and coping strategies, including insurance. Global warming, higher sea levels and more extreme weather may be a longer-term threat.

Caribbean hazards are worsened by the limited capacity of each island alone, small pools of trained personnel, sometimes poor communications and their narrowly-based economies. Specific areas of underdevelopment – from the Haiti bidonvilles to impoverished indigenous people of the Guyana rainforests – are particularly vulnerable. However, there are clear advantages in the cooperation between small islands, close contact within small populations, and the location of the region near the United States for early warning and assistance.

Caribbean disasters can be costly, especially as a proportion of Gross Domestic Product (GDP). The impact on national economies has been significant: hurricanes between 1980 and 1988 effectively reversed growth rates (figure 7.1). The Caricom Working Party on Insurance and Reinsurance estimated that Hurricanes Luis and Marilyn, which hit the Leeward Islands in August and September 1995, cost St. Kitts and Nevis 149 million US dollars, Antigua and Barbuda $254 million, and Dominica some $175 million.

In September 1989, Hurricane Hugo left ten dead on Montserrat and total damage estimated at $240 million – five years of GDP – with 98 per cent of housing damaged or destroyed. The three main hotels were out of business for at least four months, crops were ruined, the main jetty was destroyed and the fishing sector lost boats, buildings and equipment.

Jamaica suffered direct losses of $956 million from Hurricane Gilbert in 1988, including damage to agriculture, tourism, industry, housing and economic infrastructure. According to the Planning Institute of Jamaica, Gilbert was also responsible for a 30 per cent rise in inflation, a $220 million increase in government expenditure, and a rise in the public sector deficit from 2.8 per cent to 10.6 per cent of GDP.

Hurricanes David and Frederick devastated the Dominican Republic in 1979 within five days of each other, killing 2,000 people and leaving 100,000 families homeless. Health and education services suffered big setbacks.

Following the record-breaking 1995 hurricane season of 19 named storms, including 11 hurricanes, the Caricom Working Party on Insurance and Reinsurance found that the damage "was accentuated by insufficient attention to risk management and vulnerability reduction techniques...an effective approach will incorporate both legislative and voluntary elements."

From the arrival of tropical storm Arthur on June 17 to the departure of Hurricane Marco on November 25, the 1996 hurricane season proved to be long and intensely active with 13 named storms, including nine hurricanes. The International Federation of Red Cross and Red Crescent Societies launched two appeals, for Cuba and the Dominican Republic. The International Federation has supported ten Caribbean relief operations in the past two years, accentuating the need for the International Federation to develop further its own community-based disaster preparedness.

There were more than 80 major natural disasters in the Caribbean region between 1963 and 1992. Data from the Organization of American States (OAS), Department of Regional Development and Environment, indicate that from 1960 to 1994 natural disasters led to more than $9 billion losses in the 16 Caribbean countries, directly affected 9.3 million – almost half the 20.5 million total population of 1990 – and killed 14,817 people. Vulnerability to even minor disasters is accentuated by poor housing. The OAS Caribbean Disaster Mitigation Project (CDMP) estimates that 60 per cent of the housing stock is built without any technical input.

The World Meteorological Organization (WMO) has underlined that any single event causing loss of life and great destruction can reverse years of development. Escalating insurance prices and the high cost of reconstruction and rehabilitation are putting considerable strain on small island economies which depend highly on tourism and fishing.

WMO has advised that social and economic losses would be reduced by putting in place organized community-based systems to prepare for and combat their impacts. These systems must include long-term efforts in early-warning mechanisms, local preventative measures, community education, preparedness, response and rehabilitation. Even in less-developed countries, adherence to these measures reduces damage and loss of lives.

Permanent threat

In Dominica, where the 1995 hurricane season destroyed 90 per cent of the banana crop, the country's main export, the Dominica Red Cross Society, with the support of the International Federation, the European Community Humanitarian Office (ECHO) and the British Overseas Development Administration (ODA), provided food assistance to over half the island's population of 70,000.

Tourism in the Caribbean – the major employer – fell sharply after the 1995 hurricane season, with severe effects on employment. Although hurricanes can give some warning, they pose a permanent – perhaps increasing – threat to the region (see Box 7.1): 61 have made landfall in the Caribbean over the period 1899 to 1996. A bad hurricane season is likely to cause beach erosion, hotel closures, crop destruction, fishing-gear losses, deaths and injuries.

Hurricane Lili crossed Cuba with devastating effect on October 18, 1996, forcing the evacuation of 269,995 people. In the aftermath, 66,681 people lived in temporary shelters for several weeks before returning home or moving in with relatives and friends. This resulted in one of the largest Caribbean appeals in recent years, seeking 2.2 million Swiss francs to provide food, medical aid and housing materials in the affected provinces of Pinar del Rio, La Havana, Havana City, Matanzas, Villa Clara, Cienfuegos, Sancti Spiritus and Ciego de Avila, where 5,781 homes were destroyed and 78,855 damaged.

Hurricanes are just one Caribbean danger. Other hazards include volcanoes, earthquakes, floods, landslides, technological and transport accidents. Their frequency, severity and complexity – especially in combination or close sequence – reinforces the need for a highly-skilled and knowledgeable population to cope with the dangers.

Regional vulnerability to technological disasters and their impact on the economy and environment was highlighted by a 1996 cyanide spill into a major river system in Guyana. According to the Pan-American Health Organization (PAHO), from 1984 to 1987 the annual number of oil spills in Trinidad and Tobago climbed from 89 to 245. There were 13 chemical spills between 1984 and 1996 in the region, oil spills in St. Kitts, a chemical fire in Barbados, and aircraft crashes in Jamaica, Barbados, Dominican Republic, and Suriname in the past ten years. On the roads, Jamaican drivers face the third-highest fatality rate in the world, after India and Ethiopia.

But "natural" disasters – the combination of natural events with vulnerable people – have been a far bigger threat. Recent Caribbean landslides – sometimes linked to storms or floods – have forced communities to move, limited development in high-risk areas, killed more than a score of people and proved costly in repairs and debris clearing, while floods unconnected with hurricanes pose a perennial and costly problem.

Since 1889 there have been 31 serious flood events in the region. From 1979 to 1996, floods hit Jamaica every 1.5 years on average, directly affecting 420,723 residents, killing livestock and destroying infrastructure. A 1979 flood created a 600-acre lake up to 27 metres deep in the Newmarket area. Barbados suffered floods in both the 1980s – when Speightstown was flooded every year from 1984 to 1987 – and the 1990s. All other territories have experienced floods in the 1990s, with St. Lucia and Dominica facing successive events in 1994 and 1995, which disrupted communications, damaged agriculture and housing, and increased government costs.

The Caribbean plate is an active seismic area which cuts across the region and is moving west while neighbouring plates are moving east, creating two active-strike slip zones on the northern and southern borders. To the east, the North American plate is dipping under the Caribbean plate. Along this subduction zone is located the Ring of Fire, 30 or more active and potentially-active volcanoes stretching from Saba to Grenada along the 700-kilometre island chain.

Records going back centuries highlight the vulnerability of the islands to earthquakes. The biggest earthquake known to have affected the Eastern Caribbean occurred on February 8, 1848, and was felt from St. Kitts to Dominica. Deaths included more than 5,000 in Guadeloupe.

From 1950 to 1993, nine mid-level earthquakes occurred in the Caribbean: St. Kitts in 1950, St. Lucia 1953, Trinidad 1954, Jamaica 1957, Dominica 1974, Antigua and Barbuda 1974, Guadeloupe 1985, Montserrat 1985, and Jamaica 1993. None of these events could be considered major, yet the recent reclassification of eastern Jamaica from category three to category four effectively places the major population centres of Kingston and St. Andrew, Spanish Town, Portmore, sections of St. Mary, Portland, and St. Thomas in the same earthquake-risk category as Los Angeles.

At least one of the volcanoes along the Ring of Fire, the submerged Kick 'em Jenny, could threaten all its neighbours and communities as far away as Haiti and Jamaica with tsunami. Predicted wave heights vary from 0.55 to 3.55 metres in Tobago to 4.35 to 28.30 metres in Anguilla.

Volcanic activity and pyroclastic flows have paralysed the life of the 102-square-kilometre island of Montserrat since July 18, 1995. The continuing emergency emphasizes that the region is one of the most dangerous, populated, volcanic areas in the world, with at least 250,000 people living in close proximity to a possible major volcanic hazard.

There have been eruptions on several islands in this century. On May 8, 1902, a pyroclastic flow from the Mt. Pelée eruption on Martinique took 28,000 lives and destroyed the town of St. Pierre. In the same year, St. Vincent's Soufrière volcano killed 1,565 people; it last erupted in 1979.

Clouds of dust

Montserrat remains a place where the ground trembles, mountains crack and day turns to night when silica-loaded ash clouds darken the sky before falling like a silent rain. "It's ashin' again." The phrase has entered the lexicon of the tiny island. The trade winds whip ghostly vapours off the treetops and blow ash from the grass towards the sea but there is always plenty more.

The Soufriere Hills Volcano has built up a constantly-changing dome. No longer does the 3,002-foot Chance Peak dominate the skyline but its unnamed offspring, which has reached heights of at least 3,161 feet before collapsing and climbing back up again. When dome rock falls it can send up enormous clouds of dust and ash as high as 40,000 feet, producing static electricity and causing lightning that sometimes cuts the island's electricity supply.

The ash scurries under the hot canvas of military tents serving as classrooms, vexing pupils and teachers who may not have slept well in a makeshift shelter. It does nothing for the humour of the man who begs a piece of sponge for his artificial leg from a Red Cross nurse as she makes her daily rounds of evacuees.

On the west coast, the seaside capital, Plymouth, lies abandoned at the end of a long gully that could eventually channel the fury of the volcano into its deserted streets. The town is mummified in ash, buried under drifts of gritty powder which clog everything and threaten roofs with collapse. There is hardly a blade of grass left in a town famous for the beauty of its gardens.

The American University medical faculty, once the largest employer on the island, lies abandoned, surrounded by the empty houses where good money was made providing its 600 students with lodging. Also empty, the new hospital, school and government buildings only recently completed after the devastation wrought by Hurricane Hugo in 1989. The town seems destined to become a Caribbean Pompeii, ash-smothered and forgotten.

Further south in the hillside hamlet of Trials, Charles Ryner plays host in his empty bar, Strangers in Paradise, famous as a gathering place for cricket aficionados during

Box 7.1 **Forecasting a new era of hurricane-spawned destruction**

Weather forecasting is often a thankless task; hurricane forecasting seems folly. Yet that is what Professor William Gray and a team of scientists at Colorado State University Atmospheric Science Department have done for 12 years while studying Atlantic hurricanes.

Failures add important factors to the Gray prediction process. Forecast problems in 1989 led them to take Western Sahel rainfall into account, as above-average rainfall in West Africa increases hurricane activity; in 1993, they did not anticipate a resurgent El Niño, the anomalous warm eastern Pacific surface temperatures which depress Atlantic basin hurricane activity.

An underforecast for 1996 – five to seven hurricanes, compared with the actual nine – is attributed to several issues. Most years following very active hurricane seasons are below average, and 1995 had 19 named storms, including 11 hurricanes. The Quasi-Biennial Oscillation (QBO) variable east-west oscillating stratospheric winds which circle the globe near the equator was from the east in 1996; wind from the west usually indicates enhanced hurricane activity.

The scientists are now taking into account changes since 1994 in Atlantic sea surface temperatures (SSTs), with reports of decreased ice flow through the Fram Strait between Greenland and Spitzbergen causing increased salinity in the North Atlantic. Rising salinity increases water density and chilling high salinity surface water then creates dense water which can sink to great depths. The sinking water increased the flow of deep water south towards the equator and a northward flow of warm replacement water at or near the surface.

In its 1996 paper, Extended Range Forecast of Atlantic Seasonal Hurricane Activity for 1997, the Gray team show a decrease in intense hurricane activity from the late 1960s to 1994, a trend it associates with cooler North Atlantic SSTs and warmer South Atlantic SSTs in these decades.

If the opposite circulation is underway, the team anticipates a long-term increase in Western Sahel rainfall, a decrease in summer upper-tropospheric westerly winds over the tropical Atlantic and a multi-decade rise in US and Caribbean Basin intense hurricane frequency and landfall. "A new era of increased hurricane-spawned destruction appears to be approaching."

Initial 1997 predictions are for the Atlantic hurricane season to be above the 45-year (1950 to 1995) average, with seven hurricanes (45-year average: 5.7); 11 named storms (45-year average: 9.3); 55 named storm days (45-year average: 46); and 25 hurricane days (45-year average: 23). While likely to be less active than 1995 and 1996, the two worst consecutive seasons on record, 1997 is still predicted to manifest net tropical cyclone activity of about 10 per cent above average.

Was global warming to blame for recent bad years? The team says no: "Man-induced greenhouse gas warming, even if a physically-valid hypothesis, is a very slow and gradual process that, at best would be expected to bring about only small changes in global circulation over periods of 50 to 100 years, not the abrupt and dramatic one to two year upturn in hurricane activity."

Test Matches. Ryner tried a day or two in a shelter during the first evacuation in July 1995, but soon came home and refused to go during the second in December 1995, or the third in April 1996, which has yet to end. He reflects on the threat drifting over his bar. "I don't think anything will happen. This destiny what I'm confrontin' is my choice. Let me have it."

Box 7.2 **Montserrat volunteers meet the volcano challenge**

Volunteers of the Montserrat Red Cross, a branch of the British Red Cross, have been in the thick of efforts to mitigate the effects of the volcano on the 10,000 population. Alongside the government's Emergency Operations Centre, they targeted support at shelters for the elderly and disabled.

It has been a story of "displacement, evacuation, trauma and fear," according to Red Cross director Lystra Osborne, recalling preparations for the first evacuation in 1995, during the most active hurricane season in 60 years.

Volunteers helped set up two shelters, and the Red Cross provided food, cleaning items and other supplies. Then Hurricane Luis arrived with heavy rains and 220-kilometres/hour winds. For safety, some people had to be moved again while the hurricane cut off water and electricity.

Red Cross volunteers were also busy visiting the 23 other shelters housing 1,054 people, assessing needs, and distributing mattresses, commodes, food, stoves and dust masks, while also responding as Hurricanes Luis and Marilyn damaged some houses.

Working with the old and infirm, the Montserrat Red Cross decided a 50-place home for the elderly and a daycare centre was needed in the north of the island. Many across the region have helped, such as telephone giant Cable and Wireless and the Indigenous Banks of the Caribbean.

The Montserrat Red Cross, backed by the British Red Cross and the International Federation's regional delegation, also began to respond to emerging non-relief needs. Children in families facing economic hardship have been given schoolbooks and clothing, while the Jamaica Red Cross organized holidays for Montserrat schoolchildren.

The Montserrat Red Cross also worked closely with the PAHO Emergency Office in Barbados to play a full role in developing the Volcanic Crisis Mass Casualty Management Plan. Two of the most experienced first aid instructors in the region, Wayne Payne of Barbados Red Cross, and Hubert Pierre of St. Lucia Red Cross, helped train search-and-rescue teams and some 70 other key personnel, with an emphasis on treating burn injuries.

Payne, whose work coincided with the December 1995 exodus from the capital, Plymouth – giving trainees the chance to help a real evacuation – said training was important both for the skills learned and confidence.

To offer psychological support to islanders the Red Cross collaborated with the Montserrat Christian Council to facilitate stress management workshops for 120 shelter managers, teachers and householders, which were run by Montserratian psychologist, Dr Carol Tuitt.

"People were psychologically prepared in the same way as for a hurricane. They thought it would hit and then they would clean up and get on with their lives. The reality is much different," says Dr Tuitt.

Workshops suggested ideas, such as a newsletter for teachers, helped people decide whether to leave the island, and inspired other meetings. Evacuation means some families have had to split up; supportive friends and neighbours may have left. Thousands are living off the island.

Schoolteacher Dudley Meade is working in a makeshift school – mainly for children living in overcrowded shelters – at Air Studios, silenced by Hurricane Hugo but once famous for its association with top musicians, such as the Beatles, Elton John and Eric Clapton.

Some classes were in hot and dusty tents, pupils arrived tired and it was hard to concentrate. "The children would end up being sick, with runny eyes, sneezing and coughing. Children were more aggressive. You could sense the increase in stress even among the teachers."

One young displaced girl, Keisha Allen, captured her anguish in a poem. One verse went:

Home, Home, I want to go home
My Chance's Peak Mountain awaits me.
My lush green grass is crying out loud
My black sand beaches are painted with ash
And my home is falling apart
Before they all collapse in the ash
I want to go home.

Country	Average growth rate GDP 1980-1988	Average growth rate GDP 1989-1991
Dominica	4.9	4.3
Montserrat	3.7	- 4.4
St. Kitts / Nevis	6.0	4.9
Antigua / Barbuda	6.8	2.2
Jamaica	5.0	0.8

Figure 7.1
Disasters in the
Caribbean can have
a significant impact
on GDP and growth.

Of those who were willing to be relocated, 4,658 went to private homes and 1,139 were living temporarily in churches, schools or purpose-built shelters. The government's Emergency Operations Centre (EOC) initially provided food to all relocated people, as unemployment jumped from 5 to 50 per cent. In August 1996 each relocated person began getting $25-a fortnight food stamps.

The British government, through the ODA and the island government, took on most rehabilitation costs, including an immediate GB£ 8.5 million, and then £25 million after the August 1996 decision to begin a fast-track development programme for the safe area in the north of the island, where half of the population were to be relocated.

The unpredictable power of the volcano has to be measured against the natural reluctance of thousands of Montserratians to be moved. The strength of feeling is captured in the Calypso hit from island singer, Arrow, "I Just Can't Run Away". A community which has survived slavery and hurricanes over the centuries is threatened by that most insidious of natural disasters, a volcano giving no hint of its expiry date.

"There's probably nowhere else in the world where scientists have ever worked so closely with civil administrators," says the governor of this British Crown colony, Frank Savage. The scientists and the civil administration together responded to the desperation of the population and the reality that 500 people continued to live in the danger zone by producing "The Montserrat Hazard Map Microzonation."

Notes with the map capture the Montserrat dilemma. Note 1 reads: "The zones established should be regarded as delineating areas of different levels of danger, not safety. ALL of the evacuated areas remain potentially hazardous." Note 2 reads: "This map is based on the premise that not all parts of the evacuated area are equally dangerous. While this premise is self-evidently true, the area cannot be subdivided with precision: the boundaries shown are for rough guidance only."

Within weeks, all four zones identified were affected by another eruption. Around 2,000 people fled their homes as, at about midnight on September 17, 1996, the mountain triggered off an apocalyptic fireworks display. For the first time, houses were destroyed in the long-abandoned village of Long Ground. One man who had refused to leave the danger zone woke up to see a fiery rock crash through his roof.

The University of the West Indies' SRU has seized on the Montserrat crisis to point out that its annual budget from Caribbean governments of under $500,000 is grossly insufficient to monitor the risks of both volcanoes and earthquakes in the region. The SRU declared that "the net result of this low level of financial support over the decades is dilapidated monitoring and research facilities, inadequate personnel, outdated hazard maps and a public which is poorly educated on earthquakes and volcanoes." It called for $1.9 million to upgrade volcano and earthquake monitoring and research facilities, prepare seismic and volcanic hazard maps and educate people on hazards.

It added: "Monitoring and research on earthquakes and volcanoes form an integral part of the sustainable development efforts of these islands by providing relevant disaster-mitigation solutions in the shaping of development plans."

Seismic monitoring

Watching Dr William Ambeh, the seismologist who heads the SRU, expertly wielding a cutlass to clear away brush at an abandoned sugar mill overlooking the scorched Tar River Valley leaves little doubt about the courage and dedication of SRU scientists.

Cement is poured into a hole to hold a new sensor, one more to add to the nine already permanently-monitoring seismic activity. Dr Ambeh adjusts his glasses and remarks: "There's a small pyroclastic flow." A dark, frothing cloud had separated from the vapour shrouding the crumbling volcanic dome to disperse itself down the Tar River Valley, once home to some of the lushest vegetation on the island but now resembling a much-overworked quarry.

Pyroclastic flows are often dubbed the "cloud of death". Such a cloud killed 30,000 sleeping people on the island of Martinique in 1902. It is the main volcanic hazard in the East Caribbean, along with ash falls, lahars and tsunamis. A pyroclastic flow consists of rapidly-moving blasts of a mixture of hot gases, ash, fine pumice and rocks of up to 900 degrees centigrade, which pours downhill, drifting a little in the wind. On May 12, 1996, the worst one to date ran two kilometres through the Tar River Valley and out over the sea, turning the Caribbean into a boiling cauldron.

During a heart-stopping helicopter ride up the steaming, sulphurous flanks of the volcano, Dr Ambeh is philosophical about the risks vulcanologists take, and believes that the SRU is the most realistic and cost-effective way of assessing these hazards and raising attention to them. He fears that the infrequency of volcanoes and earthquakes prevents their potentially-massive impact getting the priority they deserve, and says poor SRU funding for 15 years means it could not deal as well as it should with more than one crisis. "Just coping with this one is a big task while making sure the other islands are monitored."

Dr Ambeh wants to transform the capacity of the SRU, whose origins go back to 1952. It now has 15 permanent staff, including scientists at its Montserrat Volcano Observatory and five academics based permanently at the small and dilapidated headquarters building on the St. Augustine campus in Trinidad.

The SRU wants to upgrade the existing monitoring network in the Commonwealth East Caribbean, assess the seismic hazard of all the islands and check the microzone risk of the main towns, particularly those with areas built on unconsolidated material, such as reclaimed land, where buildings could be at greater risk from shock waves. At the same time hazard maps for each island with active or potentially-active volcanoes would be upgraded.

The Montserrat Red Cross hopes to collaborate with the SRU on its community-based Disaster Preparedness Programme to educate the public about earthquakes and volcanoes in a region usually geared to the hurricane-season threat to lives and livelihoods. The SRU points out that loss reduction measures can only be successfully implemented if the public and government officials are well-informed. "Acceptance of proposed mitigation measures, many of which require an economic commitment, depends critically on the public's perception of the necessity and utility of the measures."

Unable to carry out sustained education campaigns alone across all the islands it serves, the SRU recognizes that the comprehensive network of national Red Cross societies, branches and volunteers would be an ideal vehicle to spread seismic awareness. National Disaster Coordinators would have a key role, working with other organizations, such as the Caribbean Disaster Response Agency, the US Agency for International Development (USAID) and the US Office of Foreign Disaster Assistance.

The urgent need for better education was emphasized at the Second Caribbean Conference on Natural Hazards and Disasters in Jamaica in 1996. The SRU pointed out that a number of attempts had been made to get the Montserrat public prepared for the crisis. These included a first volcanic-simulation exercise on the island in December 1988. Disaster-preparedness capability was rigorously tested when the island was hit head-on by Hurricane Hugo in 1989. However, in January 1994, only 50 people turned up when the SRU and EOC mounted a two-week poster session on earthquakes and volcanoes in Plymouth.

That urgency was brought home to scientists at the Montserrat Volcano Observatory when some teachers, who were asked to stay in classrooms with children during the less-threatening phreatic explosions of mainly steam and hot water, ran out because they thought they were pyroclastic flows.

Box 7.3 **Learning lessons to strengthen preparedness**

A survey funded by the British and Montserrat governments in 1988 warned of the volcano risks to the capital, Plymouth. That was a year before Hurricane Hugo hit Montserrat, damaging or destroying 98 per cent of homes and many other buildings. Instead of rebuilding Plymouth, its new jetty, hospital, library and government buildings, a new capital could have been developed out of the danger zone.

Mapping the Volcanic Hazards from Soufriere Hills Volcano, Montserrat, West Indies, Using an Image Processor was prepared by academics at the University of Reading's Department of Geography and the University of the West Indies' Seismic Research Unit and published in the *Journal of the Geological Society*, London.

The authors used image processing techniques to simulate future eruptions and generate risk maps, including a new type – the sequential hazard zone map – which they suggested was better than existing methods and could "be used to suggest priorities for evacuation of communities at risk at the earliest period of an eruption. It could also supply useful information on the viability of evacuation routes at different stages of eruption."

Identifying the south-west of the island as the most vulnerable to eruption, they reviewed past crises, deduced that magma was present in the volcano and warned that it posed "a considerable potential threat", adding: "With no previous experience...the Montserrat government authorities need to have a full assessment of possible hazards from the next eruption."

They accurately predicted that predominant easterly trade winds would disperse ash-fall over the south-west coast, bringing Plymouth and its environs to a halt, forecasting: "The next major eruption will probably involve the explosive emplacement of pyroclastic flows and the building of a summit dome of acid andesite", more or less what has happened since July 18, 1995.

Hurricane Hugo presented an ideal opportunity to put vulnerability reduction into practice by relocating communities at risk, and the paper was not used when creating a national disaster-preparedness plan, which – in draft during the first eruption – concentrated on hurricanes.

Many other lessons can be learned. PAHO emergency coordinator, Dr Dana van Alphen, an observer at a recent mass-casualty exercise, emphasized the need to retain skilled people, particularly in the health sector. Training first aid instructors is a 1997 Red Cross priority.

A 1996 training needs-assessment by David Sanderson, of the Oxford Centre for Disaster Studies, focused on three preparedness areas: practising mass casualty and evacuation plans; capacity building; and better public and media awareness and communication training.

But he warned: "Lessons being learnt from the current crisis need to be documented. Other islands...need to learn the many lessons being taught from Montserrat which may increase preparedness. Currently this is not happening in a way that would strengthen other islands."

Catastrophe protection

Given the combination of many hazards, widespread vulnerability and limited preparedness, the Caricom Working Party on Insurance and Reinsurance has recommended that "strong and deliberate efforts be adopted for risk management and vulnerability-reduction action by all segments of society", institutionalizing this at policy and operational levels in a national strategy with comprehensive hazard maps and appropriate materials in all education curricula.

However, big changes in attitudes will be needed before vulnerability reduction is accepted as a part of the planning process, according to 1995 paper on Caribbean catastrophe protection by the OAS and World Bank, as part of the USAID-funded Caribbean Disaster Mitigation Project.

It added: "Politicians, business leaders and individual homeowners remain with hazy and reluctant appreciations. They base their attitudes on vulnerability reduction being something for somebody else to tackle as they in the past got by without taking any deliberate actions or incurring expense...disasters `affect other people', and governments rightly step in to sort things out as do international aid organizations. To cap it all, there is nothing in it but hassle and cost."

Industry and government officials agree that underinsurance is "a substantial problem", said the paper. Many homeowners have responded to soaring insurance rates by reducing or eliminating their coverage. Absence of home insurance was one of the easiest criteria for people to meet when applying for Red Cross assistance following Hurricane Luis in September 1995. Very high insurance costs, or the simple refusal to offer cover, both removes a valuable coping mechanism for families thus increasing vulnerability now and for the future and creates new barriers to economic development in the Caribbean.

Vulnerability reduction is a three-year strategic aim for the International Federation and the 16-member national Red Cross societies in the region. Learning from recent disasters, a community-based Disaster Preparedness Programme has been developed with three components: training, logistics, and radio communications.

Building on existing materials, the training component will include developing a complete disaster-preparedness curriculum for both National Societies and vulnerable communities. The training will also be made available to other grassroots organizations. Aims include:

- improving regional and national disaster preparedness and response capacity of governments and regional agencies;

- improving National Society capacity to prepare and manage disaster plans;

- greater interaction between the local community, National Societies and other organizations;

- increasing the pool of resource people trained at national and regional level in disaster management;

- publication of a comprehensive disaster-management manual; and

- disaster-manager training modules based on the manual.

Disaster-management training targets disaster managers, and community-based disaster preparedness (CBDP) training will target vulnerable groups and areas. Both forms of training aim to improve human resources in disaster preparedness. Disaster managers will coordinate CBDP training, so the two are regarded as integrated and will run in parallel.

Part of effective response in the Caribbean lies in the rapid mobilization of relief items to meet emergency needs. International supplies are often late, so immediate response has to come from within the islands. Improvements in local emergency response will be pursued through the logistics component of the Disaster Preparedness Programme. Some improvements followed the 1995 hurricane season with two extra storage containers on Antigua and Dominica and a new warehouse funded by the Netherlands Red Cross on St. Maarten.

The final component of the programme is radio. A review has begun of the Red Cross radio communications system in the Caribbean. As well as technical improvements, this should bring better coordination and information gathering, ensuring efficient and timely disaster response. As well as being directly beneficial, the Disaster Preparedness Programme should also make a big difference by raising awareness of the need for vulnerability reduction.

Since many of the hazards in the Caribbean cannot be prevented and may in the case of hurricanes be getting worse from changing weather trends, with the long-term potential of global warming still to come, the emphasis on preparedness, early warning and vulnerability reduction is vital.

That requires investment of the time and energy of individuals, the collective endeavours of communities, resources from governments and other institutions the building of effective partnerships and commitment, which can only come from the clear understanding of carefully-assessed risks.

Chapter 7 *Sources, references and further information*

Gray, William M. et al. *Summary of 1996 Atlantic Tropical Cyclone Activity And Verification of Authors' Seasonal Prediction.*

Gray, William M. et al. *June to September Rainfall in North Africa: Verification of Our 1996 Forecasts and an Extended Range Forecast for 1997.*

Gray, William M. et al. *Extended Range Forecast of Atlantic Seasonal Hurricane Activity for 1997.* Department of Atmospheric Science, Colorado State University.

Gray, William M. et al. *Extended Range Prediction of ENSO Conditions for the Period of August 1997 to February 1998 and Verification of Last Year's Forecast.*

McCann, Janice and Shand, Betsy. *Surviving natural disasters: how to prepare for earthquakes, hurricanes, tornados, floods, wildfires, thunderstorms, blizzards, tsunamis, volcanic eruptions, & other calamities.* Salem: DIMI Press, 1995.

Organization of American States. *Working Paper on Catastrophe Protection in the Caribbean.* In collaboration with the World Bank.

Pan-American Health Organization. *Assessing needs in the health sector after floods and hurricanes.* Technical paper, no. 11, Pan American Health Organization, Washington, D.C., 1987.

Report of the Caricom Working Party on Insurance and Reinsurance, February 1996.

Web sites

PAHO: http://www.who.paho.org/

Recent earthquakes worldwide: http://www.civeng.carleton.ca/cgi-bin/quakes

Tropical storms and hurricanes worldwide, a constantly-updated site: http://www.solar.ifa.hawaii.edu/Tropical/tropical.html

Chapter

8 Mitigating and managing floods in China

Floods are a way of life in China. In a normal year, monsoon rains feed the rivers, which flow through the central and southern provinces, irrigating fields and boosting the July rice crop. Numerous reservoirs and dykes help alleviate major flood damage. Smaller floods deposit fertile silt on farmland. The alluvial plains of China are drained by seven major river systems and cover an area of over one million square kilometres. They contain half the 1.2 billion population of China and most of its important cities.

But for the rice farmers of central and southern China, high winds and floods made 1996 a year of death, destruction and disaster. One after another, typhoons from the South China Sea pounded the southern provinces and then moved inland to unleash intense rainfall. Monsoon run-off swelled the Yangtze River, which quickly filled lakes and reservoirs and then burst its banks, breaking levees and inundating large swathes of land in at least nine provinces.

Millions of soldiers and civilians were sent to reinforce dykes and load sandbags to protect cities, often to no avail. Flood-waters proved more powerful than human barriers. As summer turned into autumn and the rains continued, the Yellow River unleashed even more damage. Hundreds of thousands of people had to be removed from a strip 5 to 10 kilometres wide for hundreds of kilometres along the river. The disaster crowned six consecutive years of flooding in some of China's provinces.

Floods remained the major challenge in a year which also saw many other crises. Across 20 provinces many millions of people were affected by snow and hail storms, earthquake, fires and even drought, as well as typhoons and cyclones. Many more people have died in previous floods in China, usually because of disease or famine. Today, epidemics can be contained and food can be distributed to prevent hunger. But the duration of the rains, area of land inundated and magnitude of the damage made the 1996 floods the worst in Chinese history.

> *...official national statistics set out the destruction: 200 million people affected, 3,048 killed and 363,800 injured.*

The official national statistics for 1996, quoted in the International Federation of Red Cross and Red Crescent Societies' final report on its floods appeal, set out the destruction: 200 million people affected, 3,048 killed and 363,800 injured, 3.7 million houses/rooms destroyed and 18 million damaged, over 24 million hectares of land badly affected. Direct economic losses exceeded 100 billion yuan (12 billion US dollars).

Downpour

The annual monsoon rains in China began in late May 1996. A month later it was clear that something unusual was happening. By 2 July, the Yangtze River had risen to 33.18 metres in the central Jiangxi province, 4.68 metres above the "danger" level. In what was to become a ritual repeated across a dozen provinces over the next three months, the army was sent in to move people to higher land before flooding began or to rescue tens of thousands of villagers trapped by rising waters. Reports from southern provinces indicated that people had been killed, thousands left homeless and huge areas of farmland were swamped.

By 4 July, in Zheijiang province, several dams had collapsed and 400,000 people were surrounded by water. In villages and towns across the country, inhabitants ran for their lives after hearing the sound of the gong, banged by officials as flooding became inevitable. In some cases, people refused to leave and had to be either left behind or taken by force.

The calamity brought out both positive and negative behaviour. In Guiyang, capital of Guizhou province, taxi drivers doubled their prices. One man set up a ladder for passers-by to climb out of the flooded streets, charging 100 yuan ($12) per person. But there were many more tales of unselfish heroism and dramatic rescues. Soldiers sometimes had to swim from house to house to rescue stranded residents. Nineteen-year-old Jiang Cailian was swept away by flood-waters while searching for her family in their watermelon field in north-western Gansu province. She survived a 12-hour ordeal as the waters swept her 120 kilometres downstream, over two dams, and was finally rescued just seconds before she would have gone over the third dam.

As reports began arriving at the Beijing headquarters of the Red Cross Society of China (RCSC), money was released for emergency relief, augmenting funds that were already being raised across China by Red Cross branches and their 22 million members. On 8 July, the International Federation sent out a preliminary appeal for 3.2 million Swiss francs. On 22 July, as the disaster worsened, the appeal was increased to just over 4.4 million Swiss francs to assist 1.2 million people over three months.

Cash and goods in kind came from many governments, national Red Cross and Red Crescent societies, and the Red Cross in Hong Kong and Taiwan. The RCSC identified four principal requirements for the most vulnerable flood victims: disinfectants for sanitation and decontamination of water sources; medicine and medical supplies for treatment of water-borne diseases and first aid; food; and clothing.

The lead government agency at national, provincial and local levels was, as usual in disasters in China, the Ministry of Civil Affairs, including its Public Health Department. In cooperation with the Public Health Department, the RCSC deployed more than 60,000 medical staff in almost 6,000 medical and disease-prevention teams into the flooded areas to provide badly-needed health services.

It was an impressive operation. Each team usually combined doctors, nurses, pharmacists and epidemiologists. Given the risk of water-borne infections and the overcrowded camps of people displaced by water, a priority was to prevent outbreaks of contagious diseases. Medical teams worked round the clock in 12-hour shifts wherever flood victims had sought shelter. Most operated from tents set up in settlements; some moved around in mobile units, walking from village to village. Doctors gave first aid, while epidemiologists coordinated health monitoring, water-supply disinfection and rodent extermination.

At the height of the rains in late July, millions of soldiers and civilians were fighting the floods throughout southern and central China, including eight million in Hunan and Hubei provinces alone. The Yangtze and Yellow Rivers had already inundated large areas and were still rising. Reservoirs, dams and dykes had burst and water had reached fields that had never been flooded before. Hundreds of towns and cities were immersed in water. Millions of people were cut off, but the rains kept coming.

Humanity on a river dyke

In quick succession Frankie, Gloria and Lisa swept in: typhoons that roared toward the Chinese coast and over the mainland. They were among 41 tropical storms and typhoons to start in the Pacific in 1996. The worst to hit China were typhoon Sally and super-typhoon Herb, and among the most badly-affected areas was the region around China's biggest lake, the Dongting in Hunan.

Fed by the Yangtze and other rivers, the Dongting was like an overflowing bathtub, with gaping holes in the dykes that once held the lake back from the surrounding fields. Tens of thousands of people had managed to reach relative safety on tops of the dykes. In tents or shelters made from salvaged materials, people and animals were living under one roof, if they had a roof. At regular intervals, a Red Cross flag or a RCSC medical tent bore silent witness that the disaster victims were not alone.

In the central part of the Dongting lake area, Yiyang city would have reminded a visitor of Venice, if it had not been for the ruined houses and broken power pylons and wires lying in the water. The water had torn down dozens of dykes to reach the city, including the 889-kilometre main embankment. Residents moved along the streets on rafts made of doors, bales and planks tied together.

The local Red Cross quickly organized disaster relief, hiring trucks to distribute medicines. Over 300 medical teams were organized and hygiene messages broadcast on radio and television. Drinking water was sprayed with disinfectant. It was an expensive operation, but money was collected through public donations and a campaign which included sending 900 letters, mostly to former Yiyang residents living elsewhere.

One of the many Dongting lake villages inundated was Gong Shan Lai in Yuan Jiang county, with a population of 26,000. Late in the afternoon on 20 July, parts of the dyke protecting the village broke suddenly. Most of the people made it to safety but the first season's rice was completely destroyed. Weeks later, nearly half the village were still perched on top of the dyke in cramped and unhygienic conditions. They were exposed to the sun during the day, sweltering in temperatures of 40 degrees and more, and bitten by mosquitoes at night.

The villagers were unlikely to get off the dyke for up to two months, when the water could be drained away, and it was too late to plant for the second season crop. Like millions more, they were destined to live on food aid, if available, until the next possible crop in 1997. Half the villagers were suffering from some disease; half of those had colds, a third had diarrhoea and one in ten a skin disease.

> *Millions of people were cut off, but the rains kept coming.*

Some distance away, an entire town was under water. The long, narrow wooden boats of evacuated residents, back to find out whether their houses and belongings were still intact, flowed silently through the streets, three metres above the ground. The main street was easy to identify by the telephone poles standing out of the water. Some people had already moved back into their houses, living on the second floor and getting provisions by boat.

In Hunan province alone, the statistics tell the story of interrupted communications, hampered logistics, smashed infrastructure and increased poverty: 12,000 bridges destroyed, 8,000 kilometres of power lines and 5,000 kilometres of telephone lines down or cut, 130,000 hectares of cultivated land under water, 1.5 million homes badly damaged, and 1,000 hospitals or health centres affected.

Historic highs

By the middle of August, the floods had already reached historic highs. Over a 30-day period, 14 counties in Hubei province had experienced their heaviest-recorded rainfall. As banks and dykes burst and flood-waters covered towns and fields, 31 million people were listed as affected by the floods. On a sunny mid-August day in the village of Taiping Ko, residents were moving back into their dwellings, digging out the mud deposited by the water.

"We had no warning," said an elderly woman standing in the doorway of her half-collapsed house. "We didn't think we needed to evacuate because the government had reinforced the dykes." It was an oft-heard story. Although floods are a regular occurrence in China, inhabitants of the floodplains seem to have had enormous trust in flood-control projects, an attitude that may now be changing (see Box 8.1).

While the waters inundated towns and fields in central and southern China, a tragedy was also unfolding in Xinjiang province in north-western China. Heavy rains in the northern and southern Tien Shan mountains triggered flash floods that left 100,000 homeless as they devastated villages. Rayhen Niyaz, 40, was in her home when the flood hit, but escaped with her three boys and two girls. She and her husband had built the house themselves. "We came back to see our house and when we saw what had happened we all started to cry. We are peasants, when can we rebuild our house?"

A path of destruction was clearly visible in Fukan city. A reservoir 13 kilometres away in the hills above the city burst suddenly. A seven-metre-high wave tore through several neighbourhoods. Fifty-year-old Li Lu was in his house with his wife when the dam broke. "It all happened so suddenly. I ran this way," he said pointing toward a hill, "and my wife went the other way." Both survived, but their mud-brick house was completely demolished.

Box 8.1 **Shifting priorities for flood protection**

While Chinese authorities have been building dykes against floods for 4,000 years, there has been little information about what flood defences victims want. A recent study that attempted to measure popular opinion on the subject in the Yangtze delta led to a startling discovery: people tended to prefer non-structural remedies, such as early-warning systems, to structural ones, such as dykes.

Flood-plain agriculture in China depends on systems, sometimes centuries old, of dykes, hydraulic pumps, reservoirs and sluice gates. In a normal year, these man-made structures keep water away from the fields, except for irrigation purposes. It is only when the rains are exceptionally great that farmers have to worry about floods. The problem is, the exceptional is becoming more and more the normal.

A study reported in the journal *Disasters* of 239 villagers in two typical flood-plains of the Yangtze delta found that "contrary to expectations, the non-structural measure of flood insurance attracted more favourable responses (97 per cent) than any one of the structural measures...".

A majority of villagers did indicate a positive attitude to man-made structures to prevent floods, but many of the answers revealed a preference for flood-alleviation measures that include a combination of both structural and non-structural approaches. This is very much in line with recent Chinese policy on the flood problem.

Within the past decade and a half, the government has introduced laws on flood-risk management through flood-plain zoning, flood forecasting and warning, and flood insurance. All of these systems have become increasingly sophisticated, along with the general advancement of technology and financial resources of the society.

Flood insurance, universally popular in the Yangtze delta, aims to redistribute the losses from floods. Some 99 and 95 per cent of the inhabitants of the two areas said they were willing to purchase flood insurance. Other measures that found favour with the farmers were the regulation of flooding, building embankments and dredging.

Farmers who cultivated monsoon rice in shallow water obviously favoured regulated flooding, while others were more adamantly in favour of dykes, the higher the better. More than a quarter of the respondents said they wanted to be relocated to safer areas while very few felt the need to flood-proof their houses. Interestingly, but perhaps not surprisingly, those in medium-quality houses preferred regulated flooding while those in good-quality houses tended to prefer a complete defense against any and all inundation.

Although China bore the brunt of the unprecedented weather conditions in 1996, the typhoons claimed lives and livelihoods, destroyed homes, cropland, infrastructure and brought suffering to millions of people across Asia, from Bangladesh to DPR Korea, India to Viet Nam, many of which were only beginning to recover from the floods of 1995 and the hunger which frequently followed.

Living with floods is practically second nature to humanity. The flood-plains civilisations of the Nile, Indus, Ganges, Euphrates, Tigris and Yangtze were all sustained by the fertility of the water-soaked soil; they all used river water to irrigate their fields and all suffered periodic floods.

China has recorded more than 1,000 flood disasters since 206 BC, and for more than 2,000 years Chinese rulers have been building dykes to reach today's total of nearly 240,000 kilometres, mostly along the lower reaches of major rivers. At the same time, such man-made structures have all too frequently been overwhelmed by nature.

In 1880, five million people died from drowning, disease and starvation. In 1920, 200,000 people died: half from drowning, the rest from starvation or exposure. Floods and droughts were and are prime causes of food scarcity in China. From 1850 to 1932, studies suggest starvation killed 5 per cent of the affected population, while 13 per cent emigrated.

Some floods have even been caused intentionally. Chiang Kai-Shek blew a hole in the levees of the Yellow River in 1938 to stem the advance of Japanese troops. One million people died; the Japanese were delayed a few weeks. Another Chinese general, Gao Mingheng, had used the same strategy to suppress a peasant rebellion in 1642.

Humans at the centre

Disease control, better early warning and improved relief can all save lives, while urbanization, deforestation, soil erosion, migration, landlessness and poverty all increase vulnerability by either encouraging more floods or putting more people in their way. While river floods may simply exceed the channel carrying capacity and overrun any defences, humans may soon be causing more floods and high-wind disasters through global warming.

In a warmer world, there will be more frequent hurricanes, cyclones and tornadoes. Most "general circulation models" for the impact of global warming indicate higher temperatures, greater evaporation and thus increased rainfall.

The Intergovernmental Panel for Climate Change (IPCC) warned in 1996: "Warmer temperatures will lead to a more vigorous hydrological cycle; this translates into prospects for more severe droughts and/or floods in some places, and less severe droughts and/or floods in other places. Several models indicate an increase in precipitation intensity, suggesting a possibility for more extreme rainfall events."

Worryingly for a country such as China, where tens of millions are already on the move for work, it adds that the gravest effects of climate change could come "through sudden human migration, as millions are displaced by shoreline erosions, river and coastal flooding, or severe drought," and says that 70 million Chinese are among many millions more worldwide threatened by sea-level rise.

> *70 million Chinese are among many millions more worldwide threatened by sea-level rise.*

Disturbances in the hydrological cycle would have far wider impact. One-third of the world's fresh water is currently controlled in one way or another, by dams, reservoirs, locks and sluices, according to the study, *Water; The International Crisis*, by Robin Clarke, published in association with the Swedish Red Cross. All of these structures were built to cope with existing levels of water and past predictions of extreme events.

More than a decade ago, the World Meteorological Organization warned that important economic and social decisions about major water-resource management activities – irrigation, hydropower, drought relief, agricultural land use, structural designs, coastal engineering, energy planning – were being based "on the assumption that past climatic data, without modification, are a reliable guide to the future. This is no longer a good assumption."

Building defences

With or without global warming, floods cause enormous damage and this damage is increasing every year. Some people deliberately risk living in flood-prone areas, since these are usually areas with high fertility from the natural hydrological cycle. Others have no choice but to return after each disaster to the same cyclone-prone coastal zone. Flood damage is likely to increase as population grows and urban areas expand, forcing the most vulnerable on to marginal land prone to flooding, or hillsides at risk of flash floods or mudslides. Damage that was previously caused by a flood with a "return period" – the predicted time between extreme events – of 100 years can now be caused by a 20-year flood.

The very act of building defences against flooding can encourage people to live in flood-prone areas. Man has yet to build a structure which nature cannot destroy. The cost of safeguarding buildings and fields against a flood of a size that is anticipated only once every 100 years is often so prohibitive that governments shy away from it. But even insufficient defences can inspire misplaced confidence with terrible consequences. The feeling of security has, in many parts of the world, led to intensive development at the expense of shrinking forests and eroded pastures, both of which add to the disaster risk.

Misperceptions increase the risk of such unwarranted development. Traditionally, the greatest danger has been from large rivers pouring enormous amounts of water over large areas during bad floods. Now, major flood disasters tend to occur in densely-populated areas in small river catchments.

Not everyone loses in floods. As one writer on Bangladesh found: "All day I had been travelling across flooded rice fields, and in my mind that added up to catastrophe, disaster. But to Rajendra (a local fisherman) the flood meant good things: the chance to use his boat to visit neighbouring villages, rather than walking all day; an abundance of fish; and sediment left behind that makes the land bountiful and gives him the chance to grow his own rice. `Such water,' he said, `what the Lord has given us.'"

Felling forests, spurring floods

As increasing human construction downstream results in greater damage from floods, the floods themselves are also becoming bigger and more frequent because of human activity upstream. Deforestation and more intensive use of pastures are destroying the soil's capacity to hold water and slow its path to rivers.

Rapid deforestation along the upper reaches of the Yangtze and Yellow Rivers have cut the time between rainfall and drainage into the river systems. The steelmaking campaign urged by Mao Zedong during the "Great Leap Forward" in 1958 fostered the building of small wood-fueled, iron-smelting furnaces, encouraging the wholesale felling of forests. Later another campaign, "Grain First", led to further deforestation as woods had to make way for fields. Vegetation cover in China dropped from 19 per cent in 1949 to 13.3 per cent today.

...floods are not only a curse but also a blessing.

The Chinese government is well aware of the problem and has set large reforestation plans for the vital Sichuan province. But experience shows that plans are not always fulfilled and people who plant trees for the benefit of those living downstream may not find it in their interest to keep the trees. One government report admits that of the trees planted since 1949, only one-third have survived.

In Asia, as elsewhere, floods are not only a curse but also a blessing. While threatening lives and livelihoods, they also bring life to the floodplains by depositing fertile silt on rice paddies and other farmland. The Yellow River derives its name from the yellow loess soil which may make up 40 per cent of the volume of its "water". This "soil soup" annually floods an area of 8,300 square kilometres, depositing the silt along its 4,000-kilometre journey to the sea.

The bed of the river is continuously rising and, to prevent floods, dykes along the side of the river have to be built up at the same rate, and need constant, costly maintenance. Even when dykes do work, they can do more harm than good. They have a negative impact on the fertility of the surrounding fields, forcing farmers to use other fertilizers to maintain yields. A dyke upriver keeps the water in its channel, causing more of it to go downriver and threaten residents there.

Flood-prevention channelization also causes increasing riverbed silting, requiring ever higher dykes in a never-ending vicious circle. The surface of the Yellow River runs eight metres above the surrounding fields for much of its way through the Chinese countryside. Despite all efforts for hundreds of years, the river – known as China's Sorrow – has killed more people than any other feature of the earth's surface.

In order to avoid the curse but retain the blessing, many have urged greater attention to non-structural rather than structural methods of flood-alleviation. In essence, structural measures tend to emphasize flood-prevention – typically by keeping rivers within their channels – while non-structural measures give priority to saving lives and property without blocking periodic inundation.

> *It seems clear that floods will continue to become greater, more frequent and more destructive.*

Flood forecasting is the best-known, non-structural, flood control measure. Authorities in China have developed modern and sophisticated warning systems. Certainly, many more people would have perished in 1996 had these systems not been in place. The fact that the major rivers originate and flow within China – unlike countries where shared rivers bring tension and limited cooperation – greatly eases the task of warning people downriver of impending inundation.

Considerable resources have gone into expanding and improving flood forecasting and warning in the flood-prone countries of Asia. But those resources are dwarfed by the money that is spent on physical structures. Since 1949, for example, China has spent $27 billion, 4 per cent of all its public outlay, on water-control projects.

It seems clear that floods will continue to become greater, more frequent and more destructive. Global warming is likely to continue, deforestation is far from being checked and population pressures will continue to put more property in the way of the water. At the same time, man-made structures are unlikely to protect against big floods and may even be part of the problem rather than the solution.

Technological advances in flood forecasting and warning systems, however, and increased emphasis on such non-structural measures, may serve to reduce the danger from flooding. Disaster preparedness – including the use of RCSC Disaster Preparedness Centres (the last of which was completed in 1996) built by the RCSC with multi-donor backing through the International Federation – greater attention to disease control, quick relief action, and well-organized rehabilitation after the disaster will all help minimize the dangers posed to human lives.

Eventually, man may find that many of the ways in which he has been fighting floods have been futile. Better technology and different priorities may make it possible to live with – instead of protecting against – this deadly natural disaster.

Chapter 8 *Sources, references and further information*

Alexander, David. *Natural Disasters.* UCL Press Limited, 1993.

Chang, William Y.B. *Effects of water management policy on the formation of Lakes Hongtze and Kaoyu, Two Large Chinese Lakes.* Journal of Great Lakes Research, 1993, Volume 19, Number 4.

Clarke, Robin. *Water; The International Crisis.* London: Earthscan Publications Ltd., 1991, in association with the Swedish Red Cross.

Cobb, Jr, Charles E. *Bangladesh: When the Water Comes.* National Geographic, June 1993

Dudgeon, David. *River Regulation in Southern China: Ecological Implications, Conservation and Environmental Management.* Regulated Rivers: Research and Management, Vol II, pp. 35-54, 1995.

Ives, Jack. *Floods in Bangladesh: who is to blame.* New Scientist, 13 April 1991.

Legome, Eric et al (Ohio State University). *Injuries Associated with Floods: The Need for an International Reporting Scheme.* Disasters, Volume 19, Number 1.

Mushtaque, A. et al. *The Bangladesh Cyclone of 1991: Why So Many People Died.* Disasters, Volume 17, Number 4.

Pearce, Fred. *The Dammed; Rivers, Dams and the Coming World Water Crisis.* London: The Bodley Head, 1992.

Rasid, Harun et al. *Structural vs Non-structural Flood-alleviation Measures in the Yangtze Delta: A Pilot Survey of Floodplain Residents' Preferences.* Disasters, Volume 20, Number 2.

Topping, Audrey R. *Ecological Roulette: Damming the Yangtze.* Foreign Affairs, September/October 1995.

Wickramanayake, Ebel. *Flood Mitigation Problems in Vietnam.* Disasters, Volume 18, Number 1.

Wijkman, Anders and Timberlake, Lloyd. *Natural Disasters: Acts of God or acts of Man?* London: Earthscan Publications Ltd., 1984.

World Meteorological Organization. *Hydrology of Disasters.* Proceedings of the Technical Conference, Geneva, November 1988.

Worldwide Fund For Nature. *Climate Change due to the Greenhouse Effect and its Implications for China.* Gland, 1992.

Web sites

EarthAction: Global Warming no time to lose: http://www.oneworld.org/earthaction/global_warming.html

Greenpeace: "Special Report: Climate Change and River Flooding": http://www.greenpeace.org

International Federation (China situation reports): http://www.ifrc.org

International Rivers Network: http://www.irn.org/links/sites.html

Reliability of climate models: http://gcrio.gcrio.org:80/CONSEQUENCES/fall95/mod.html

US Global Change Research Program: http://www.usgcrp.gov/usgcrp/CCSOKNOW.html

Chapter

9 Old diseases and new epidemics in the NIS

Infectious diseases, far from being eliminated as a public health problem, remain the leading cause of death worldwide. A wide range of factors – from environment, technology and mobility to greater poverty and poorer health surveillance – has revived old diseases and spread new infections; the result will be more epidemic emergencies.

Lessons from tackling disease crises, such as the sudden re-emergence of diphtheria in the Newly Independent States (NIS) of the former Soviet Union, suggest new attention must be paid to the threat of epidemics in a world in which so many health systems face serious economic constraints, and that strategic alliances between many different groups and institutions will be crucial if these diseases are to be contained.

In the past two decades, at least 30 new disease-causing organisms have been identified, including HIV, hepatitis C and Ebola virus. New infections emerge in different ways, through evolution of existing organisms, their spread into new geographic areas or population groups, and previously-unrecognized infections appear in areas undergoing ecological changes which increase human exposure to sources of new or unusual infectious agents.

Re-emergence of infectious diseases may occur because of the development of resistance in existing agents (e.g., gonorrhoea, malaria, pneumococcal disease) or

*Figure 9.1
Epidemics
in 1996:
International
Federation
support.*

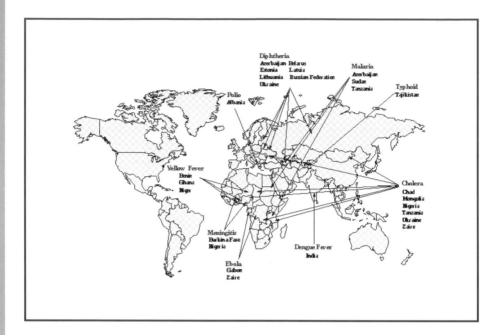

breakdown in public health measures for previously-controlled infections (e.g., cholera, tuberculosis (TB), diphtheria).

Infections from contaminated water supplies and foods place entire communities at risk, from deaths as far apart as the United States, Japan and Scotland caused by the bacterial pathogen known as E-coli to the 400,000 people made ill – and 4,400 hospitalized – in the 1993 pollution of municipal water supplies in Milwaukee, Wisconsin (USA), by the intestinal parasite Cryptosporidium. Many smaller outbreaks occur worldwide; numbers affected are increasing.

Cholera, smallpox, and plague are historical examples of infectious diseases that spread globally with devastating impact, often during periods of rapid economic change or population growth. In modern times, travel and commerce have fostered the global spread of HIV/AIDS and influenza, as well as the re-emergence of cholera as a global health threat.

Recent disease developments have convinced public health professionals that:

■ Health-policy priorities must include surveillance, investigation and outbreak monitoring.

■ Optimal health practice requires integration of epidemiology and laboratory science.

■ Communication and community-based learning will be a big factor in good disease control.

■ Improved surveillance and control systems need better health infrastructure at every level.

Encouraging this analysis have been the growing health problems of the NIS, where past systems offering state-funded guarantees of unlimited, free medical care in reality faced constraints in both finance and management, and lacked incentives to improve the quality, efficiency, and effectiveness of the health-care system.

By 1993, it had became clear that economic and political events were undermining the health system in the NIS by leading to lower living standards, poorer health care facilities, and a growing range of social problems. Adult male life-expectancy fell from 69 to 59 years and infant vaccination rates for TB, pertussis and other diseases declined, in part because the NIS had only limited hard currency supplies to buy vaccines from the Russian Federation – supply centre for the ex-Soviet Union – or elsewhere.

> *After decades of widespread immunization, diphtheria had been a disappearing, almost forgotten, disease.*

After decades of widespread immunization, diphtheria had been a disappearing, almost forgotten, disease. Its swift re-emergence was aided by large population movements following the relaxation of travel controls and a rise in conflicts, as well as social and economic stress.

Decisive action

A key factor was low immunization rates, especially as vaccine-induced immunity wanes in time without boosters. In some areas, less than 50 per cent of adults were fully protected. Use of lower-dosage adult vaccine on children put many at greater risk, and adherence to an overlong contraindication list against immunization left significant numbers unprotected.

The generally low awareness of diphtheria, and limited health-sector capacity to respond because few health professionals had personal experience of the disease, may also have contributed to the delay in taking decisive action to control the epidemic.

As the World Bank *Development Report* has highlighted, investment in primary childhood vaccination is one of the most cost-effective public health measures. Faced with the disease and vaccination situation in the NIS, donor discussions led to the creation of the International Immunization Coordination Committee (IICC) in 1994.

The Japanese government agreed to support the secretariat and two biennial meetings for the international consortium, whose members now include the European Community (EC), the US Agency for International Development (USAID) and smaller non-governmental organizations (NGOs). Control of the fast-expanding

Country	Jan.-March 95	Jan.-March 96	Change
Estonia	7	4	- 43 per cent
Latvia	107	31	- 71 per cent
Lithuania	13	4	- 69 per cent
Belarus	82	44	- 46 per cent
Ukraine	1364	800*	- 41 per cent*

*preliminary unofficial data

Figure 9.2
Diphtheria cases in 1995 and 1996.

Box 9.1 Reaching the hard-to-reach for screening and immunization

The changes in the former Soviet Union have been accompanied by increased poverty, addiction and unemployment. Hard-to-reach groups, including the homeless, are both vulnerable to diseases, such as diphtheria, and more difficult to assist.

One example of successful efforts to prevent the spread of diphtheria was a social mobilization and vaccination campaign by the local Red Cross – backed by the International Federation and other National Societies – to help the homeless in Petrozavodsk in Karelia Oblast, close to the Finnish border.

Police helped identify likely locations, and volunteers and visiting nurses contacted homeless people to talk about health issues and give out a leaflet about diphtheria, its transmission and prevention, as well as details about vaccination sessions.

After hard work to overcome shortages of basic equipment, such as gloves, syringes and needles for blood samples – even hospitals keep these items for emergencies only – sessions consisted of a medical check-up, screening for infectious diseases and diphtheria vaccination.

A mobile X-ray machine was set up, and smear and blood samples were taken from each person to test for TB and HIV.

During the physical examination, those being vaccinated were screened for lice and treated, if necessary, with a special shampoo.

To encourage attendance, tea, sandwiches and second-hand clothes – donations from the Swedish Red Cross – were given out. If screening showed any positive result, the homeless person was immediately treated or referred to specialists.

In just the first two days of the campaign, more than 80 homeless people were vaccinated against diphtheria and screened for other diseases. As well as fighting diphtheria epidemics, this approach builds awareness of communicable diseases among disadvantaged people.

Despite limited government support, low public profile, and the small number of staff and volunteers, the Red Cross in Karelia performed well and gained useful experience, giving it more confidence for future activities.

diphtheria epidemic in all 15 republics of the former Soviet Union was one of the most important issues on the IICC agenda.

The diphtheria epidemic had broken out in the Russian Federation in 1990, moved on to Ukraine, and soon spread to all other NIS. Figures collated by the IICC and the World Health Organization (WHO) show that there were 839 cases in 1989 in all the NIS, but by 1993 the figure had reached 19,504 (of which 15,209 were in Russia), and, by 1994, cases had almost peaked at 47,628 (39,582 in Russia). Despite a fall in Russian cases to 35,652 in 1995, the NIS total for the year reached 50,425.

The epidemic threatened to spill beyond the NIS, with a small number of imported cases in Bulgaria, Finland, Germany, Norway and Poland in 1994, and epidemiological projections suggested that without aggressive measures, 1995 and 1996 would see 340,000 cases in the NIS. Only mass immunization could slow, and finally control, the epidemic.

In June 1995, the International Federation of Red Cross and Red Crescent Societies launched a joint appeal with the UN Children's Fund (UNICEF) and WHO for 33 million US dollars to contain and, if possible, eliminate the diphtheria epidemic in all the countries of the former Soviet Union.

Alongside the appeal, the IICC assisted development of a joint strategy for member governments and all supporting partners, as well as country visits and necessary information to other donor agencies.

An important positive factor was that individual, designated members of the IICC from donor nations were also managers of country-specific support programmes, ensuring vital personnel continuity. The ensuing networking process gave credibility to the overall strategy and to national implementation plans.

Under the strategy with WHO and UNICEF, the International Federation took operational responsibility in support of government efforts for Belarus, Ukraine and the Baltic states of Estonia, Latvia and Lithuania, a total population of approximately 72 million. It has been a major provider of vaccines, syringes and needles, cold-chain equipment, antibiotics, antitoxin and human resources to implement and monitor the programme, and has encouraged social mobilization efforts by national Red Cross societies.

An estimated 79 million doses – including a margin for wastage – would be required to provide diphtheria vaccination to all people not fully immunized. Under the responsibility of the International Federation in 1995 and 1996 more than 39 million doses of vaccines were shipped into the five countries. Timed to match the main infection periods – the return to school in September, and the period from January to March – mass vaccination campaigns took place in autumn 1995, and in spring and autumn 1996. The case load and mortality fell dramatically.

> *Only mass immunization could slow, and finally control, the epidemic.*

Under the International Federation umbrella and with funds from the European Community Humanitarian Office (ECHO) and other donors, the Finnish Red Cross implemented campaigns in Ukraine, and the German Red Cross assisted campaigns in Belarus in 1995. In the Baltic states, ECHO funds were supplemented by Japan and many European countries.

By the end of 1996, almost 150,000 people had been infected and 4,000 killed, but the epidemic was under control in the Baltic states and Belarus. After almost universal, steep rises in 1993-1994 and 1994-1995, comparing the 1995 peak year with 1996 shows sharp falls in Belarus from 322 cases to 179, Estonia from 19 to 14, Latvia 369 to 112, and Lithuania 43 to 11. In Ukraine the situation is not as positive. Despite efforts to control the epidemic and a fall in cases from 5,280 in 1995 to 3,156 in 1996, high incidence and mortality rates have persisted; mass vaccination campaigns for selected oblasts (regions) have continued.

Overall, the 1996 figures for all NIS showed 20,215 cases, a fall from 1995 of more than 30,000. Russian cases were down from 35,652 to 13,604, and figures for all other NIS had more than halved to 6,611. More work will be needed to finally control this re-emerging disease. It will be of particular importance to improve clinical management, laboratory diagnosis and epidemiological understanding of "high risk" and "hard-to-reach" groups.

New partnerships

National Red Cross societies were key partners in the diphtheria campaign, but faced a simple challenge: most had never been so closely involved in disease-outbreak control programmes before. There were also some early uncertainties about the differing roles of government and National Society in the new context of market economics and pluralist politics.

It soon became evident that national Red Cross societies were more than capable in this new field of medical relief, managing a wide range of social mobilization, communication and outreach efforts: training seminars for health staff; printing and distributing information materials about diphtheria and vaccination sessions; and involving national and local media in often innovative ways throughout the programme.

Another key area was building new partnerships to encourage community-based and private institutions, such as schools, commercial companies, the police and other

Box 9.2 **Offering relief health education over the Internet**

Health management during the Great Lakes crisis has resulted in an enormous improvement in aid agency understanding of health-care requirements for large numbers of refugees and displaced people. This new knowledge is being gradually integrated into public health teaching, while a large number of research questions still remain unanswered.

This combination has created a new job opportunities for health workers in public health, particularly in issues related to the health of unstable populations. Most public health professionals in mid-career have neither the time nor the money to attain necessary qualifications in the traditional way, but many would be willing to undertake "distance learning" while remaining in existing posts.

New opportunities for long-distance learning are offered through present-day electronic media. A first experience with a Masters degree course at the School of Public Health, John Hopkins University, in the United States has been successful.

The Relief Health Department of the International Federation has decided to develop a masters course in international public health with John Hopkins University, to be taught on the Internet. This is the first course in relief health of its kind worldwide, and will be advertised on the Web.

A detailed plan is being developed for content, certification and case studies from relief operations, including Great Lakes and Somalia, for the teaching. The project will be guided by the expressed needs of returning health delegates, the results of programme evaluations, and emerging operational needs. Emphasis will be given to health management issues, the inclusion of development goals in the early phase of emergency response, and the increasing documentation of "what works" in this new field.

Given the high mobility and speed required for success in disaster management, the global approach to "virtual learning" follows what has been achieved in communication elsewhere.

groups, to spread information to the public. In this field, the churches were very important routes to spread important health messages.

Poor and vulnerable people are more affected by infectious diseases than those with higher incomes, so priority must be given to control epidemics among hard-to-reach groups, such as the jobless, homeless, alcoholics, and travelling peoples.

National Red Cross societies rose to this challenge, using their widespread networks of volunteers in the branches, who visited market places and dormitories to boost community knowledge, and ensured – through home-visiting nurses or urging neighbours to help those unable to travel – that many house-bound and disabled people were vaccinated at their homes.

Simple techniques to offer welfare assistance – organizing soup kitchens for the vulnerable, distributing second-hand clothes – were also employed as valuable self-selecting mechanisms to help identify those in greatest need and offer them vaccination.

Although infection and death rates have declined, Red Cross volunteers remain vigilant and have begun operations in areas where vaccination was incomplete, such as remote villages. Participation in the diphtheria campaign has increased their capacity enormously, raising confidence for other programmes, such as TB control in hard-to-reach groups.

Learning from the diphtheria campaign, all the NIS continue to face three closely-related problems in vaccine-preventable disease control:

■ lack of preventive maintenance in the health sector;

■ continuity of basic vaccination services; and

■ control of the current diphtheria epidemic.

Box 9.3 Meningitis in Africa: tracking effectiveness and efficiency

Major problems in disease surveillance, vaccine supply, and immunization evaluation showed up in 1996 during the largest epidemic of meningococcal meningitis which affected around 250,000 people in 17 countries of sub-Saharan Africa and killed at least 25,000.

At least $15 million were spent immunizing up to 25 million people, but there is no way of determining if the campaign was effective, or whether the money and effort were totally wasted.

Though low when compared to malaria, tuberculosis and gastroenteritis, the case load and mortality figures are high, given that both a fully-effective vaccine and antibiotic therapy are available. The incidence was two or three times that of the previous epidemic in the "meningitis belt" in 1989.

The number of cases in the region had begun to rise in 1995, when Niger reported 43,200 cases. Attack rates increased to 350 per 100,000 population in Burkina Faso in 1996. The population of the affected countries is about 200 million, with an average attack rate of approximately 100 per 100,000.

In total, some 20 to 25 million vaccine doses were purchased by national governments, bilateral donors, WHO, UNICEF and the International Federation. In Nigeria alone, 10 to 15 million doses were used, and one million in Chad. In some districts 40 to 50 per cent of the population was covered, and in some urban areas the coverage rate was reportedly 60 per cent.

Patchy reporting and variable exposure to the disease make it difficult to determine if the vaccine campaign had any effect in decreasing the frequency of the disease. This is complicated by the fact that modelling of vaccine effects in African epidemics shows at best a 50 per cent reduction of disease in settings where the vaccine is used most efficiently.

Major problems affected vaccine supply in 1996, including forecasting, procurement, production capacity, quality and financing. Predicting the development of the epidemic was difficult, since reporting depended on many systems of uncertain accuracy in separate administrative regions, so that incipient outbreaks were not spotted, or identified too late.

In addition, even though previous patterns of epidemics were known, national forecasts for vaccine requirements were not made. If national forecasting had been done, it would have shown that 70 to 120 million doses would be needed from February-March 1996, representing 80 per cent coverage,

plus a small wastage factor, for a population of 80 to 140 million people.

There was no coordination or careful quantifying of the overall need, and some requests were filled in duplicate. WHO and others supplied vaccine on a first come, first served basis. Sporadic and overlapping requests prevented the true demands being known, which stopped the manufacturers making early decisions to rapidly produce larger quantities.

In some areas the vaccine arrived after the peak of the epidemic curve, covered just 40 per cent of the population, and could therefore only have limited effects.

Pasteur-Merieux, the main vaccine supplier, was able to meet all commitments despite the repeated requests, but no one manufacturer could have met a demand for 70 million doses within a few months.

The quality of vaccine was high and met all international standards at the production site. More difficult was vaccine quality at the injection site, due to uncontrolled transportation and storage procedures. In addition, a counterfeit vaccine was distributed by unscrupulous dealers in some areas.

One dose of vaccine cost 18 US cents; $4.5 million were therefore required for the campaign. Adding the cost of transport, storage, delivery, and needles and syringes brought the total to $15 to $20 million – a very large expenditure for poor countries to achieve an effect on human disease.

Major problems arose due to the inadequate amount of reserve vaccine for immediate use, and the lack of coordinated identification of areas which should have had priority because of need and the opportunity to make effective use of supplies.

This was in contrast to the epidemics of polio in Albania and diphtheria in Eastern Europe, where the supply of vaccine was rapidly and efficiently managed by a coordination team. Both of these epidemics occurred amid efforts to provide routine preventive immunization.

As the meningitis epidemic showed, disease control can be extremely costly if economic and social factors undermine epidemic surveillance. For example, in 33 African countries, yellow fever has increased to an annual total of 200,000 cases. One million doses of yellow-fever vaccine were distributed by WHO, among others, in a successful campaign during the epidemic in Liberia and Sierra Leone. There is now a five-year, $190million comprehensive control programme for this region.

Donor assistance will be needed for several years to help the NIS meet the challenge of developing self-sufficiency in immunization, from the vaccination service itself to ensuring guaranteed supplies of vaccines meeting internationally-accepted standards. Self-sufficiency does not necessarily mean local production, but it will require local funding.

Managing the diphtheria epidemic has led to new confidence among donors about vaccine-procurement capabilities in emergencies. This can buffer the development of vaccine capacity in the NIS, giving time to develop national quality-control authorities. Technical assistance in this process can be provided by external donors, coordinated by WHO.

One technical problem being tackled is developing an indicator to show at a glance if the vaccine has been frozen, since diphtheria vaccine is damaged by freezing, a common problem in the cold climate of the NIS. This is unlikely to be introduced immediately; it took ten years to develop and introduce the heat-sensitive label used on some oral polio vaccines.

> *Infectious diseases seem set to remain the biggest killer for years to come.*

In addition to overall fast-rising rates of immunization, especially among children, two other positive moves have helped: low-potency vaccines are no longer being used for children and all countries have moved to a more limited list of contraindications against immunization.

Diphtheria control still requires intensified efforts by national health authorities, and some countries will also require sustained assistance from international organizations to achieve the crucial goal of the highest-possible immunization coverage for the whole population.

Mass mobilization

All agencies collaborated in developing the strategy in 1995 and witnessed the improvement of national plans that brought the epidemic under control. A mutual learning process included accepting different management systems, but also developing new modes of urgent interventions. The enthusiasm this created is showing first results in the polio-eradication campaign, new approaches to hepatitis B and E control, and new plans for measles eradication in 15 to 20 years.

Working alongside WHO and UNICEF, the International Federation and national Red Cross societies broke new ground in tackling a disease-control programme of this scale and scope, and have together shown their capacity for mass mobilization and ability to reach marginal people. There is a new and positive relationship between the governmental sector, the Red Cross and NGOs, and a willingness to increase professionalism in disease response and to take on new programmes.

The confidence that success encourages has also been accompanied in national Red Cross societies by a higher technical understanding of disease control, an expectation that this work is clearly within the Red Cross remit, and a new sensibility about building strategic alliances.

Lessons from the diphtheria campaign will be essential in developing an effective strategy for dealing with future health challenges in the former Soviet Union, especially the growth of HIV/AIDS, which has the added complexity of being an infection without vaccine that combines a medical crisis with issues of drug abuse, sexual behaviour and human rights.

Box 9.4 HIV/AIDS: the new disaster in the NIS?

HIV/AIDS is fast adding itself to the health problems of the NIS, according to figures from health ministries and the UN programme on AIDS.

Drug abuse is helping expand its infections rapidly in communities that were hardly affected by the epidemic only a few years ago. Unsafe sex will allow infections to multiply further.

In 1995, Ukraine saw dramatic increases in newly-infected users of injected drugs in cities bordering the Black Sea because of shared needle use and linked to a new opium-based drug. For example, the percentage of HIV-infected people among those injecting drugs in Nikolayev went from 1.7 per cent in January 1995 to 56.5 per cent in December.

In Ukraine, prior to 1995, 70 per cent of those infected had contracted the virus through sexual contact and 5.5 per cent from intravenous drug use. But, in 1995, 68.5 per cent were infected through shared syringes and 21.2 percent were infected through sexual contact; 13 per cent of all those injecting drugs in Ukraine were HIV-positive.

That switch may well be underway in Russia where, in 1995, 64 per cent of those HIV-positive were infected through sexual contact, and 1.6 per cent from injected drug use. In the Russian Federation in 1994, 84,377 tests were carried out among intravenous drug users and no HIV-positive individuals were found; in 1996, 190 of the 45,507 tests done were positive for the virus.

In 1996, the reported number of HIV-infected people in Kaliningrad increased 18-fold, from 21 to at least 387 – mainly intravenous-drug users. Following a trend observed elsewhere, the NIS male-female ratio among HIV-infected people has begun to equalize as homosexual contact ceases to be the main infection route. Recent health surveillance shows sharp increases in the rates of sexually-transmitted diseases, which indicate an increase in unsafe sex which would allow HIV infections to rise.

New syphilis cases per 100,000 people rose in the Russian Federation from 81.7 in 1994 to 172 in 1995, in Belarus from 72.1 to 147.1, in Moldova from 116.6 to 173.6, and in Kazakhstan from 32.6 to 123.

NIS health specialists recently toured US hospitals treating AIDS patients, expressing their concern about HIV-related spread of drug-resistant TB, and the high costs of drug therapy. They took back ideas for needle exchanges, an HIV-information hotline for health workers, out-patient AIDS clinics, and peer education programmes by HIV-positive volunteers.

In the US, 650,000 to 900,00 people are infected with HIV. Final verdict of the visitors? "The situation with AIDS in the United States is basically our future," said Dr Alexander Goliusov, special AIDS adviser to the Russian Federation Minister of Health and chief specialist in the Department of Preventative Medicine in the Ministry of Health. Goliusov oversees the 82 AIDS centres throughout Russia, the first of which opened in 1989.

Worldwide, infectious diseases seem set to remain the biggest killer for years to come, and other major epidemics, whether of "old" or "new" diseases, are inevitable, given the erosion of health systems, antibiotic resistance, disruption from war, and the threat of climate change.

Population movement itself increases the risks: every day, one million people cross an international border; every week, a million travel between the South and North; in 1996 over 33 million people were either refugees or internally displaced, while scores of millions are on the move in search of work; and every year, half a billion people travel by air.

While humanitarian agencies may be tempted to see health problems as just a subset of other disasters, the scale and speed of recent epidemics – from cholera around Goma to diphtheria in the NIS – suggest these could be key humanitarian challenges of their own, requiring investment to enhance capacity and knowledge, and new ways of working closely with other groups globally and locally, often with limited or no government assistance on the ground.

A positive development has been the close collaboration between technical staff across UN agencies, NGOs and the International Federation. Without formal mechanisms and despite institutional differences, these links have allowed useful and effective cooperation.

In a far more formal move, WHO is revising the International Health Regulations, which allow the United Nations (UN) and governments to control the movement of people and goods to block cross-border epidemics. The regulations at present only cover cholera, plague and yellow fever, and are being updated to match the scale of global traffic and trade in the 21st century.

As new questions come forward – food safety, genetic engineering, animal parts in human transplants, resistance from antibiotic misuse – to be added to known concerns – poverty, environment, overcrowding – the economic and political context of diseases will be crucial.

> *Companies cannot see profits in researching vaccines for the diseases of poor countries and poor people.*

The changes in the NIS are a dramatic example of a clear trend in health care worldwide towards privatization of provision, from the failed or frayed states of Africa and others undergoing structural adjustment to the market-focused countries of Asia and the Americas, and the shrinkage of welfare systems in developed economies.

As the former Soviet Union demonstrates, such changes can cause a steep decline in health standards, especially among the poor, old, disabled, homeless or jobless. Where systems collapse, private health care is often unaffordable or unobtainable; where insurance-based systems take over, these will both take a high percentage of income from those with least and be unable to extend cover to those in greatest need, increasing vulnerability.

Since no national or international institution – aid agencies, churches, charities – can sustainably substitute for the state, health care offers an indication of the limits of government retreat and private sector advance. Where the state is no longer the main provider of health care, a far wider set of agents – local authorities, companies, NGOs, national Red Cross and Red Crescent societies, the UN, churches and community groups – must come together in strategic alliances to tackle health crises, such as epidemics.

Cuts in research

A different sort of problem is already emerging as governments cut research funding. Companies cannot see profits in researching vaccines for the diseases of poor countries and poor people, and the supply of ideas for countering microbial threats is beginning to slow.

Yet re-emerging infections and evolving diseases can affect people anywhere, regardless of lifestyle, culture, ethnicity or status. The Nobel Laureate geneticist, Dr Joshua Lederberg of Rockefeller University, has said: "The microbe that felled one child in a distant continent yesterday can reach yours today and seed a global pandemic tomorrow."

Less than ten years after the US military first supplied its field physicians with penicillin, it was Lederberg who demonstrated that natural selection operated in the bacterial world to create flourishing drug resistant strains. Today he has been quoted characterizing the solutions to disease emergence and re-emergence as international in scope, multitudinous, largely straightforward and commonsensical. But he adds: "The bad news is, they will cost money."

Chapter 9 *Sources, reference and further information*

Berkelmann, R.L. et al. *Infectious Disease Control: A Crumbling Foundation.* Science, No. 264, pp. 368-370, 1994.

Centers for Disease Control and Prevention (CDC), International Federation, WHO, UNICEF. *A WHO/UNICEF Strategy for the Control of Diphtheria in the European Region.* Berlin, December 1995.

CDC. *Addressing Emerging Infectious Disease Threats.* US Department of Health and Human Services, Atlanta, Georgia, 1993.

Cohen, Felissa L. *Emerging infectious diseases: nursing responses.* Nursing Outlook, July 1996.

Galazka, A.M. et al. *Resurgence of Diphtheria.* Europ.J.Epidem., No. 11, pp. 95-105, 1995.

Garrett, Laurie. *The coming plague: newly emerging diseases in a world out of balance.* New York: Viking Penguin, 1995.

Henig, Robin M. *A dancing matrix: how science confronts emerging viruses.* New York: Random House, Inc., 1994.

Horowitz, Leonard G. *Emerging viruses: AIDS & Ebola - nature, accident or intentional?* Rockport: Tetrahedron, Inc., 1996.

Lederberg, J. et al (eds.). *E-merging Infections. Microbial Threats to Health in the United States.* Institute of Medicine, Washington DC, 1992.

Morse, Stephen S. (ed.). *E-merging Viruses.* Oxford, New York: Oxford University Press, 1993.

Slavkin, Harold C. *Emerging and re-emerging infectious diseases: a biological evolutionary drama.* Journal of the American Dental Association, January 1997.

WHO. *New and re-emerging infectious diseases.* World Medical Journal, January 1996.

Web sites

Centers for Disease Control: http://www.cdc.gov

International Federation of Red Cross and Red Crescent Societies: http://www.ifrc.org/

WHO: http://www.who.ch

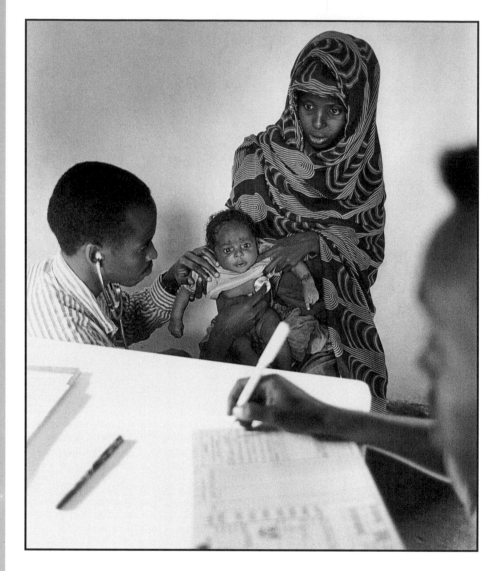

Falling funds for assistance are not matched by falling disaster numbers. Global statistics continue to highlight the gap between opportunity and achievement. But as priorities for government spending shift, will we lose vital information systems and key data?
Somalia, 1994.

Chapter

10 Facts and figures for effective response

Data on disaster occurrence, its effect upon people and its cost to countries remain, at best, patchy. The *World Disasters Report 1997* draws upon five main sources of information for the data presented here: the Centre for Research on the Epidemiology of Disasters (CRED); the US Committee for Refugees (USCR); the Department of Peace and Conflict Research (DPCR) of Uppsala University, Sweden; the Organisation for Economic Co-operation and Development's (OECD) Development Assistance Committee (DAC); and INTERFAIS, a World Food Programme (WFP) information system. Each organization is described in more detail below.

One of the key problems today with disaster data collection is the lack of standard, accepted definitions. In the data from Uppsala, for instance, a major armed conflict is defined as one where there are more than 1,000 battle-related deaths. Thus, many civil war situations do not appear in this section. Likewise, problems exist over such loose categories as "internally-displaced" people or even people "affected" by disaster.

> *Relative changes and trends are more useful to look at than absolute, isolated figures.*

Most of the data in this chapter, except that on government humanitarian sending, is culled from a variety of public sources: newspapers, insurance reports, aid agencies, etc. The original information is not specifically gathered for statistical purposes and so, inevitably, even where the compiling organization applies strict definitions for disaster events and parameters, the original suppliers of the information may not.

The figures therefore should be regarded as indicative. Relative changes and trends are more useful to look at than absolute, isolated figures.

CRED

Based at the Department of Public Health, Catholic University of Louvain (Belgium), CRED has developed a system of databases for global disaster management, drawing on its existing disaster documentation, information network and computer system.

Tables 1 to 12 of this database section were derived from the EM-DAT, a disaster-events database developed by CRED and sponsored by the International Federation of Red Cross and Red Crescent Societies, World Health Organization (WHO), United Nations Department for Humanitarian Affairs (UNDHA), European Community Humanitarian Office (ECHO) and the International Decade for Natural Disaster Reduction. USAID's Office of Foreign Disaster Assistance also collaborated in getting this database started.

EM-DAT is now fully operational, with more than 10,500 records of disaster events from 1900 onward, and its own menu for updates, modification and retrieval. Designed to have the right level of detail for wide use, the entries are constantly reviewed for redundancies, inconsistencies and the completion of missing data.

USCR

The US Committee for Refugees, which supplied the data for Tables 13, 14 and 15 of this chapter, is a non-governmental information and advocacy programme. Its goal is to produce positive change in the lives of the world's uprooted people, and is affiliated with the Immigration and Refugee Services of America. It is concerned primarily with conditions affecting refugees, internally-displaced people, and asylum seekers.

Since 1958, USCR has travelled to the scene of refugee emergencies to document protection, assistance, and human rights issues. It communicates concerns directly to governments, provides first-hand assessments to the press, briefs members of the United Nations (UN) and non-governmental organization (NGO) community, testifies before the US Congress, and conducts educational outreach in the United States.

In recent years, USCR has collaborated with the Representative of the UN Secretary-General on Internally Displaced Persons to draw attention to the situation of displaced people throughout the world.

USCR publishes the annual *World Refugee Survey*, the monthly *Refugee Reports,* and reports on issues of topical importance. USCR is an active advocate on asylum issues in the United States, the need for continued US funding for overseas assistance programmes, and other issues of concern to refugee advocates in the United States and throughout the world.

Department of Peace and Conflict Research

Data for Tables 16, 17 and 18 were supplied by the Department of Peace and Conflict Research of Uppsala University, Sweden. The Department was established in 1971 to conduct research and offer courses in peace and conflict studies. Research activities focus on two specific areas: the origins and dynamics of conflict, and conflict resolution and international security issues.

The former section studies the causes of war, and the diffusion of conflicts. Among the causes of conflict investigated are environmental destruction, domestic problems, military capabilities, as is the escalation of ethnic conflict. Data on armed conflicts are continuously collected and statistics on major armed conflicts have been published in the SIPRI Yearbook since 1988. Since 1993 a list of all armed conflicts appears in the *Journal of Peace Research.* The Department publishes its own annual reports, *States in Armed Conflict.* The data constitute an important source for research, and for information on the state of the world.

The second area of research deals with durable conflict resolution, for instance in border issues and ethnic conflicts. The relationship between conflict resolution, state-building and international integration is also an issue. The situations for minority populations are of interest. A particular focus has been on South-East Asia. UN conflict-resolution experiences are analysed, as are Swedish and Nordic security affairs, military expenditure and conversion of military resources, as well as general studies concerning military research and development, and arms exports.

The Department has also been active in promoting peace research as an academic discipline around the world.

DAC/OECD

The data in Table 19 have been supplied by the OECD's Development Assistance Committee, the principal body through which OECD deals with issues related to cooperation with developing countries. The Committee is concerned with support for efforts in developing countries to strengthen local capacities and thus to pursue integrated development strategies, and studies, among other topics, the financial aspects of development assistance, statistical problems, aid evaluation, and women in development.

INTERFAIS/WFP

The source of information for Table 20 is INTERFAIS (the International Food Aid Information System), a system funded by WFP. It is a dynamic system, which involves the interaction of all users, represented by donor governments, international organizations, NGOs, recipient countries and WFP country offices. All information is cross-checked before being disseminated. Its comprehensive and integrated database allows the monitoring of food-aid allocations and shipments for the purpose of improving food-aid management, coordination and statistical analysis. The database is updated on a continuing basis, and data can therefore change as allocation plans and delivery schedules are subject to modifications. Data is available from 1988.

Disaster data

Information systems have improved vastly in the last 25 years and statistical data as a result are much more easily available. An increase in the number of disaster victims, for example, does not necessarily mean that disasters, or their impact, are increasing, but may simply be a reflection of better reporting. However, the lack of systematic and standardized data collection from disasters, man-made or natural, in the past is now revealing itself as a major weakness for any developmental planning. Cost-benefit analyses, impact analyses of disasters or rationalization of preventive actions are severely compromised by inavailability and inaccuracy of data or even field methods for collection. Fortunately, as result of increased pressures for accountability from various sources, many donor and development agencies have started placing priority on data collection and its methodologies.

CRED's data – Tables 1 to 11 – have appeared in each edition of the *World Disasters Report* since the pilot issue in 1993. The economic impact of natural disasters (Table 12) – also based on data from CRED – appears for the first time in this Report. Despite efforts to verify, cross-check and review data, its quality can only be as good as the reporting system. If field agencies were to adhere to standard reporting methods, it would be far easier – and much less costly – to assemble essential data from numerous sources. Data presented in these tables are recorded at CRED; while no responsibility can be taken for a figure, its source can always be provided.

> *Many donor and development agencies have started placing priority on data collection and its methodologies.*

The criteria for entry of an event is ten deaths, and/or 100 affected, and/or an appeal for assistance. In cases of conflicting information, priority is given to data from governments of affected countries, followed by UNDHA, and then the US Office for Foreign Disaster Assistance. Agreement between any two of these sources takes precedence over the third. This priority is not a reflexion on the quality or value of the data, but the recognition that most reporting sources have vested interests, and figures may be affected by socio-political considerations.

Dates can be a source of ambiguity. The declared date for a famine, for example, is both necessary and meaningless – famines do not occur on a single day. In such cases, the date that the appropriate body declares an official emergency has been used.

Figures for those "killed" in disasters should include all confirmed dead, and all missing and presumed dead. Frequently, in the immediate aftermath of a disaster, the number of "missing" is not included, but it may be added later. Without international standards, definitions vary from source to source, so each entry is checked for clarification.

People "injured" covers those with physical injury, trauma or illness requiring medical treatment as a direct result of disaster. First aid and other care by volunteers or medical personnel is often the main form of treatment provided at the disaster site, but it has not been defined whether people receiving these services should be included as "injured".

"Homeless" is defined as the number of people needing immediate assistance with shelter. Discrepancies may arise when source figures refer to either individuals or families. Average family sizes for the disaster region are used to reach consistent figures referring to individuals.

Defining "people affected" is extremely arduous. Figures will always rely on estimates, as there are many different standards, especially in major famines and in the complex disasters of the former Soviet Union and Eastern Europe.

In this chapter, both the numbers of people affected, and those made homeless, are given by disaster. However, to obtain a more realistic idea of the numbers involved, the number of homeless for the last five years should be added to the number of people affected.

Disparities in reporting units can create dilemmas, such as the monetary value of damages expressed in either US dollars or local currencies. While it is easier to leave currencies as they are reported and convert them only when the event is of interest, this procedure effectively slows the comparison and computations often required by data users.

> *At present, estimating the monetary value of disasters is far from precise.*

In addition, inflation and currency fluctuation are not taken into account when calculating disaster-related damages. At present, estimating the monetary value of disasters is far from precise. Multi-standard reporting makes estimations difficult, as does the lack of standardization of estimate components: for example, one estimation may include only damage to livestock, crops and infrastructure, while another may also include the cost of human lives lost. It is not always clear whether estimations are based on the cost of replacement or on the original value. Insurance figures, while using standard methodology, include only those assets that have been insured, which in most developing countries represent a minor proportion of the losses. A standard methodology for the estimation of the economic damages is urgently required to justify prevention and preparedness programmes.

Disaster information from four different sources is reported regularly to CRED, and the register updated daily. The sources are: UNDHA situation reports; International Federation situation reports; *Lloyd's Casualty Week* (published by Lloyd's of London with information on weather events, earthquakes, volcanic eruptions and different types of accidents worldwide); and some World-Wide Web pages on the Internet which provide daily news on disasters in specific regions (for example, the Pan-American Health Organization page).

Each new event is entered with date, type of disaster and country. Data on human or economic impact are consolidated at CRED at three-month intervals the first year. Annual updating is undertaken the following year.

Information is cross-checked each year, and occasionally new events added, using data from reinsurance companies, the World Meteorological Organization, the UN Economic and Social Commission for Asia and the Pacific, as well as articles in specialized journals and unpublished university research. Finally, the annual list of disasters is sent for control to a focal point in countries affected by the disasters.

Changes in national boundaries can cause ambiguities in the data, most notably the break-up of the Soviet Union and Yugoslavia, and the unification of Germany. In such cases, no attempt has been made to retrospectively desegregate or combine data. Statistics are presented for the country as it existed at the time the data were recorded.

The figures in Table 12 reflect the proportion of annual gross national product (GNP) lost over the last ten years due to disaster-related damage. When insufficient data are available to assess the economic impact, the estimated damage per person affected (in US dollars) is calculated by dividing the number of people affected by a disaster with the sum of total damages for a similar disaster in the same country. The number of people affected in the disaster are then multiplied by the damage "value" per person to arrive at an estimation of the economic impact, which is then used to compute the indicator as a proportion of GNP. The average estimated damage is calculated for each disaster affecting a country over a one-year period; the sum of all important disasters will give the annual average estimated damage which is then presented as a percentage of annual GNP.

This method gives fairly satisfactory results. However, it is important to standardize data collection on an international level, as the validity of the indicators is determined by the quality of data presented.

Major armed conflicts

DPCR's data, presented in tables 16, 17 and 18, are based on publicly-available material: newspapers, journals, research reports, government sources, etc., and, if possible, on at least two independent sources. Evaluations are made by the DPCR and outside experts and, in exceptional cases, DPCR makes its own estimates of, for example, battle-related deaths, based on information about the intensity of fighting, weapons used, etc. All data are revised annually as new information becomes available.

A major armed conflict is defined as:

■ A conflict where a government is an actor on one side of the conflict, facing an organized armed opponent (either another government or an organized faction). The data therefore covers states of organized, not spontaneous, violence, in situations where two actors are in conflict, i.e., massacres by one armed faction against unarmed civilians are not included.

■ A conflict where at least 1,000 battle-related deaths have been recorded during the course of the conflict. This means that only conflicts approaching a common notion of "war" or "protracted conflict" are included. There are many minor conflicts in addition to those reported in this chapter. The battle-related deaths include those that are targeted victims of the fighting, whether civilians or military, but do not include victims of other effects of the conflict, e.g., starvation as a result of the breakdown of society.

■ A political conflict concerning government and/or territorial control. Conflicts between criminal groups are not included, as the purpose is clearly only economic gain, nor are strikes or other social troubles, unless they come to demand the replacement of incumbent government.

(In the following tables, some totals may not correspond due to rounding.)

Figures 1 to 5   Mortality from all disasters (1971-1995)

Across all the continents, the number of people killed by disaster has gone up over the past 25 years. While in Africa and Asia this partly reflects a general increase in population and hence people exposed to disaster, in Europe the upward trend reflects both the rise in traffic-accident deaths, and the wars and civil unrest of the last decade.
Source: CRED

Figure 10.1

Smoothed time-trends in disaster-related mortality in Africa from 1971 to 1995.

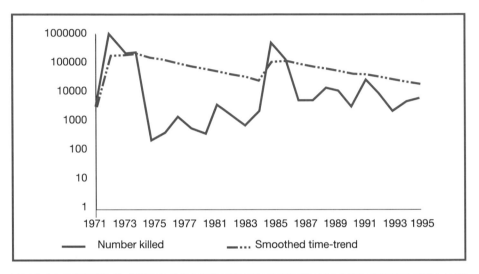

Figure 10.2

Smoothed time-trends in disaster-related mortality in the Americas from 1971 to 1995.

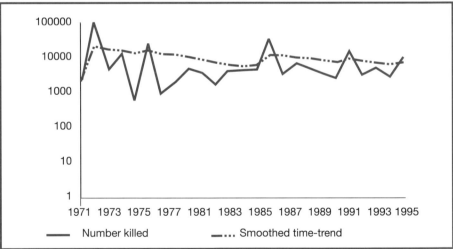

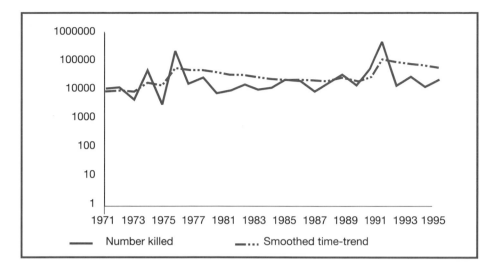

Figure 10.3

Smoothed
time-trends in
disaster-related
mortality in Asia
from 1971 to 1995.

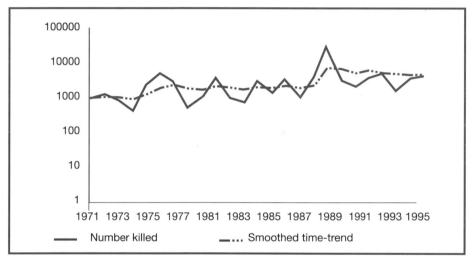

Figure 10.4

Smoothed
time-trends in
disaster-related
mortality in Europe
from 1971 to 1995.

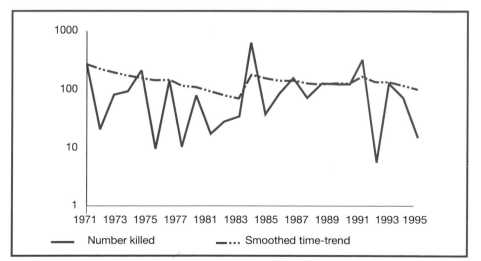

Figure 10.5

Smoothed
time-trends in
disaster-related
mortality in Oceania
from 1971 to 1995.

Tables 1 and 2 Human impact by region. Annual average over 25 years (1971-1995)

Disasters with a natural trigger

	AFRICA	AMERICAS	ASIA	EUROPE	OCEANIA	TOTAL
Killed	76,640	5,779	43,805	2,282	93	**128,597**
Injured	1,039	9,445	75,590	3,408	135	**89,616**
Affected	11,765,777	3,801,124	119,770,719	580,098	653,830	**136,571,546**
Homeless	279,837	320,181	4,082,290	71,299	14,077	**4,767,685**
Total	**12,123,293**	**4,136,528**	**123,972,403**	**657,087**	**668,134**	**141,557,445**

Disasters with a non-natural trigger

	AFRICA	AMERICAS	ASIA	EUROPE	OCEANIA	TOTAL
Killed	647	3,806	2,453	947	21	**7,873**
Injured	271	1,083	5,820	562	486	**8,223**
Affected	3,506	49,639	43,139	9,204	11,410	**116,898**
Homeless	2,526	1,691	6,869	7,784	64	**18,935**
Total	**13,106**	**55,360**	**54,287**	**17,195**	**11,982**	**151,929**

> *The average number of people affected by disasters went up six per cent in 1996 as compared with 1995's figures. Asia shows the greatest total number affected, reflecting the devastating effects of flooding (see Chapter 8). Despite its lower population than Asia, Africa remains the continent with the most disaster-related deaths, illustrating the tragic effect of civil strife and famine (see Chapter 6). Source: CRED*

Tables 3 and 4 Human impact by type. Annual average over 25 years (1971-1995)

Disasters with a natural trigger

	EARTHQUAKE	DROUGHT & FAMINE	FLOOD	HIGH WIND	LANDSLIDE	VOLCANO	TOTAL
Killed	19,051	73,621	12,732	16,075	786	1,017	**123,281**
Injured	26,333	0	20,705	9,975	250	284	**57,547**
Affected	1,688,327	60,692,579	60,041,340	11,130,893	137,608	94,325	**133,785,071**
Homeless	237,562	22,720	3,241,365	1,142,656	107,519	15,144	**4,766,965**
Total	**1,971,272**	**60,788,920**	**63,316,142**	**12,299,598**	**246,163**	**110,769**	**138,732,864**

Disasters with a non-natural trigger

	ACCIDENT	TECHNOLOGICAL ACCIDENT	FIRES	TOTAL
Killed	3,908	616	3,349	**7,873**
Injured	1,868	5,593	762	**8,223**
Affected	17,504	53,934	45,461	**116,898**
Homeless	976	8,717	9,242	**18,935**
Total	**24,256**	**68,860**	**58,814**	**151,929**

> *Famine remains the biggest killer and, in the 1996 figures, also affected the greatest number of people. Flooding, affecting similar numbers, kills fewer people, as they are able to move out of danger. Compared with these massive disasters, earthquakes and volcanoes affect relatively few people but kill a far greater proportion of those affected (see Chapter 7). Source: CRED*

Tables 5 and 6 Number of events by global region and type over 25 years (1971-1995)

Disasters with a natural trigger

	AFRICA	AMERICAS	ASIA	EUROPE	OCEANIA	TOTAL
Earthquake	41	135	252	165	85	678
Drought & Famine	296	53	88	16	16	469
Flood	184	382	653	154	135	1,508
Landslide	12	90	99	21	10	232
High Wind	84	454	685	228	199	1,650
Volcano	9	33	46	16	6	110
Other	205	99	189	94	6	593
Total	831	1,246	2,012	694	457	5,240

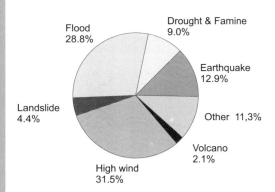

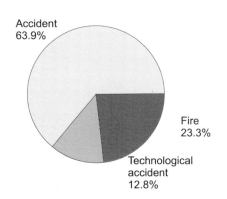

Disasters with a non-natural trigger

	AFRICA	AMERICAS	ASIA	EUROPE	OCEANIA	TOTAL
Accident	247	350	761	339	22	1,719
Technological accident	25	104	110	100	4	343
Fire	37	124	244	192	31	628
Total	309	578	1,115	631	57	2,690

Floods and high winds which often cause flooding, remain the most frequently-occurring disaster, particularly in the Americas and Asia. In Africa, drought and famine continue to dominate, accounting for one-third of the disaster events.Source: CRED

Tables 7 and 8 Total number of disasters by global region and type in 1996

Disasters with a natural trigger

	AFRICA	AMERICAS	ASIA	EUROPE	OCEANIA	TOTAL
Earthquake	0	4	6	2	0	12
Drought & Famine	3	1	0	0	0	4
Flood	14	17	23	9	2	65
Landslide	1	5	4	2	2	14
High Wind	2	16	22	2	1	43
Volcano	0	2	0	1	1	4
Other	18	1	16	3	0	38
Total	38	46	71	19	6	180

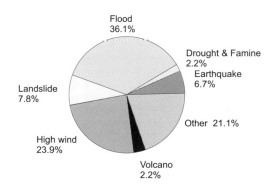

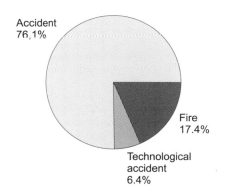

Disasters with a non-natural trigger

	AFRICA	AMERICAS	ASIA	EUROPE	OCEANIA	TOTAL
Accident	15	22	25	18	3	83
Technological accident	n.a.	3	2	2	n.a.	7
Fire	n.a.	3	10	6	n.a.	19
Total	15	28	37	26	3	109

> *The figures for 1996 alone show a 15 per cent decrease in the number of disasters globally, compared with 1995, but this figure has to be treated with caution as it says nothing about the severity of the disasters. Significantly, the number of non-natural disasters, including industrial accidents and major transport accidents, is up by 18 per cent on 1995. More non-natural disasters highlight the need for increased vigilance in work and traffic safety practices. Source: CRED*

Table 9 Annual average number of people reported killed or affected by disaster by country over 25 years (1971-1995)

Country	Killed	Affected	Country	Killed	Affected	Country	Killed	Affected
			Gambia	8	29,400	Suriname	7	n.a.
AFRICA			Côte d'Ivoire	8	286	Bahamas	4	n.a.
Ethiopia	48,448	2,716,352	Togo	8	26,865	Paraguay	3	29,623
Sudan	6,115	981,061	Sao Tomé/Prin.	7	7,483	Dominica	2	3,720
Mozambique	4,547	1,255,159	Mauritius	7	39,526	St. Lucia	2	2,944
Somalia	906	37,343	Namibia	5	16,528	Martinique	0	860
Nigeria	470	138,418	Reunion	5	6,728	Anguilla	1	n.a.
Chad	285	296,619	Gabon	4	405	Bermuda	1	n.a.
Cameroon	142	40,749	Cen.African Rep.	3	651	Uruguay	1	948
Algeria	141	28,108	Comoros	3	15,418	Trinidad/Tobago	0	2,000
Angola	138	128,488	Cape Verde Is.	3	476	Belize	0	3,731
Egypt	121	7,666	Lesotho	2	47,280	Guadeloupe	0	3,000
Zambia	119	154,029	Eritrea	1	253	St-Vincent	0	1,726
Kenya	109	441,723	Eq. Guinea	1	13	Barbados	0	n.a.
Mauritania	97	271,147	**TOTAL**	**62,632**	**10,624,083**	Antigua	0	5,600
South Africa	94	275,006				St Martin	0	1,600
Zaire	75	32,944	**AMERICAS**			St Kitts	0	72
Mali	74	209,027	Nicaragua	3,348	60,204	Montserrat	0	240
Tanzania	72	140,669	Colombia	1,161	39,911	**TOTAL**	**9,583**	**3,850,762**
Niger	71	315,725	Guatemala	979	156,860			
Malawi	69	579,561	Honduras	717	52,592	**ASIA**		
Madagascar	60	272,749	Peru	689	388,437	Bangladesh	31,870	10,867,802
Burkina Faso	55	278,845	Mexico	635	99,836	China, P. Rep.	12,847	29,423,116
Guinea	39	2,204	USA	570	49,069	India	4,728	63,723,255
Uganda	34	59,405	Brazil	418	1,478,595	Iran	2,958	71,870
Zimbabwe	28	384,210	Ecuador	178	52,064	Philippines	2,159	2,299,494
Swaziland	26	65,649	Haiti	170	219,861	Indonesia	650	319,140
Sierra Leone	25	520	El Salvador	122	65,093	Pakistan	575	1,023,233
Tunisia	24	7,567	Dominican Rep.	86	102,628	Afghanistan	489	73,311
Liberia	22	129	Chile	63	165,673	Nepal	365	251,819
Congo	20	n.a.	Venezuela	61	6,259	Viet Nam	316	1,480,076
Morocco	19	7,560	Canada	57	20,379	Japan	245	141,403
Benin	18	126,568	Caribbean	57	2,000	Korea, Rep. of	174	66,732
Senegal	17	291,256	Bolivia	50	168,117	Thailand	158	538,204
Ghana	16	521,041	Puerto Rico	45	160	Burma	107	241,067
Burundi	15	288	Argentina	43	499,251	Saudi Arabia	95	n.a.
Libyan A.J.	12	n.a.	Guyana	36	10,867	Yemen Arab Rep.	86	121,000
Rwanda	12	164,145	Cuba	34	67,904	Sri Lanka	73	558,508
Guinea Bissau	11	839	Jamaica	19	54,187	Taiwan	50	1,321
Djibouti	11	32,521	Panama	12	6,965	Iraq	50	20,006
Botswana	8	167,478	Costa Rica	9	27,786	Korea, Dem. Rep.	42	1,778

Country	Killed	Affected
Hong Kong	32	1,703
Laos	32	191,960
Cambodia	31	41,616
Yemen, P. D. Rep.	30	31,680
Malaysia	28	14,905
Maldives	9	462
Lebanon	7	2,060
United Arab Emir.	6	n.a.
Mongolia	6	4,000
Oman	5	200
Bahrain	5	n.a.
Israel	4	16
Syria	4	5,367
Bhutan	2	2,623
Jordan	1	749
Singapore	1	n.a.
Macao	0	80
TOTAL	**46,258**	**119,813,858**

EUROPE

Country	Killed	Affected
Soviet Union	1,461	61,512
Russian Federat.	850	17,161
Tajikistan	544	21,275
Turkey	517	32,869
Estonia	227	35
Italy	193	75,588
Spain	148	32,924
Georgia	109	26,500
France	88	33,412
Romania	87	58,230
United Kingdom	83	558
Greece	67	29,072

Country	Killed	Affected
Ukraine	66	102,333
Yugoslavia	53	15,154
Kyrgyzstan	42	48,750
Finland	39	n.a.
Serbia	35	n.a.
Poland	32	856
Uzbekistan	24	12,500
Portugal	24	1,607
Norway	23	n.a.
Armenia	23	52,000
Azerbaijan	21	n.a.
Germany, F. Rep.	18	4,220
Belgium	17	65
Denmark	17	4
Ireland	17	140
Albania	16	141,400
Kazakhstan	13	7,500
Hungary	12	n.a.
Moldova	12	6,250
German Dem. Rep.	12	600
Sweden	10	n.a.
Switzerland	10	290
Azores	10	852
Macedonia	8	400
Czechoslovakia	8	4
Bulgaria	6	n.a.
Austria	4	n.a.
Netherlands	3	480
Iceland	0	208
Turkmenistan	0	75
Belarus	0	10,000

Country	Killed	Affected
TOTAL	**3,229**	**589,302**

OCEANIA

Country	Killed	Affected
Papua New Guinea	41	13,024
Australia	39	573,364
Solomon Islands	15	8,887
Fiji	10	45,598
Vanuatu	3	6,244
New Zealand	2	1,958
Samoa	1	n.a.
Western Samoa	0	10,280
Kiribati	0	42
French Polynesia	0	200
Tonga	0	5,192
Tuvalu	0	40
Wallis & Futuna	0	180
Cook Islands	0	80
New Caledonia	0	80
Tokelau	0	68
TOTAL	**113**	**665,308**

Countries are ranked here, continent by continent, according to absolute numbers of people killed, shown as an average over the past 25 years. The figures exclude direct war casualties. Africa is dominated by the famine deaths in Ethiopia and Sudan. In Asia, the devastating effect of floods in clearly shown by the high death figures for Bangladesh and China (see Chapter 8). Within Europe, the Soviet Union is still shown, as data from the 1970s and 1980s cannot be desegregated by the present new states. Source: CRED

Tables 10 and 11 Average estimated damage by region and by type over five years (1991-1995) in thousands US dollars

Disasters with a natural trigger

	AFRICA	AMERICAS	ASIA	EUROPE	OCEANIA	TOTAL
Earthquake	281,200	26,835,660	108,681,322	372,500	255,000	**136,425,682**
Drought & Famine	98,739	1,860,000	82,755	1,188,600	1,008,400	**4,238,494**
Flood	240,474	19,674,365	98,518,290	84,222,045	11,100	**202,666,274**
Landslide	n.a.	25,400	222,200	60,100	n.a.	**307,700**
High Wind	150,565	54,705,154	29,672,016	1,796,960	1,361,800	**622,288**
Volcano	n.a.	10,000	212,888	n.a.	400,000	**87,686,495**
Other	n.a.	4,345,151	950,000	2,103,299	3,036,300	**7,398,450**
Total	**770,978**	**107,455,730**	**238,339,471**	**89,743,504**	**6,072,600**	**439,345,983**

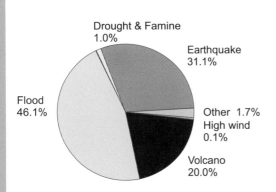

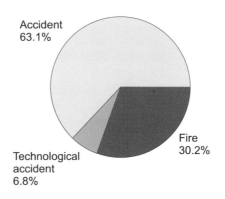

Disasters with a non-natural trigger

	AFRICA	AMERICAS	ASIA	EUROPE	OCEANIA	TOTAL
Accident	1,602,700	8,778,903	15,999,499	1,064,600	12,000	**27,457,702**
Technological accident	27,500	521,056	275,117	2,095,500	38,000	**2,957,172**
Fire	n.a.	3,635,500	6,658,865	2,688,100	150,000	**13,132,465**
Total	**1,630,200**	**12,935,458**	**22,933,481**	**5,848,200**	**200,000**	**43,547,339**

> *Data on the financial cost of disaster is extremely difficult to gather and verify, although information systems are getting better and more rigorous (see Chapter 3). The average cost of natural disasters over the past 25 years now stands at over $87 billion a year, a massive figure compared with the $3 billion figure devoted to humanitarian response globally. Increases in the 1996 average figures over 1995 reflect the factoring in the cost of China's and the DPR Korea's floods. Source: CRED*

TABLE 12 Estimation of the economic impact of natural disasters

Country	Area	GNP* ($, 1990)	Average 1985-1995	% of GNP
Benin	Africa	1,700,000,000	622	0.000037
Madagascar	Africa	2,700,000,000	17,876	0.000662
Tanzania	Africa	2,800,000,000	877	0.000031
Argentina	America	76,500,000,000	1,135.153	0.001484
Bolivia	America	4,500,000,000	24,044	0.000534
Brazil	America	402,800,000,000	551,425	0.000137
Canada	America	543,000,000,000	12,667	0.000002
Chile	America	25,500,000,000	175,658	0.000689
Colombia	America	40,800,000,000	276,303	0.000677
Cuba**	America	18,300,000,000	29,912	0.000163
Ecuador	America	10,100,000,000	78,365	0.000176
El Salvador	America	5,800,000,000	103,000	0.001176
Honduras	America	3,000,000,000	24,874	0.000829
Jamaica	America	3,600,000,000	137,000	0.003806
Nicaragua**	America	1,421,000,000	58,900	0.004145
Panama	America	4,400,000,000	7,235	0.000164
Peru	America	25,100,000,000	34,533	0.000138
Mexico	America	214,500,000,000	5,459,799	0.002545
Costa Rica	America	5,300,000,000	62,061	0.001171
Afghanistan**	Asia	5,870,000,000	36,178	0.000616
Korea, DPR	Asia	not identified	276,154	not identified
Bangladesh	Asia	22,600,000,000	385.321	0.001705
Burma	Asia	not identified	58,542	not identified
China	Asia	415,900,000,000	3,865,888	0.000930
Japan	Asia	3,141,000,000,000	50,131	0.000002
Korea, Rep. of	Asia	231,100,000,000	104,450	0.000045
Indonesia	Asia	101,200,000,000	56,344	0.000056
Nepal	Asia	3,3000,000,000	41,061	0.001244
Philippines	Asia	44,400,000,000	557,995	0.001268
Thailand	Asia	79,000,000,000	631,233	0.000799

Viet Nam**	Asia	11,997,000,000	30,876	0.000257
India	Asia	294,800,000,000	1,645,507	0.000558
Iran	Asia	139,100,000,000	377,167	0.000271
Pakistan	Asia	42,600,000,000	162,227	0.000381
Sri Lanka	Asia	8,000,000,000	109,058	0.001363
Yemen Arab R.	Asia	not identified	6,300	not identified
Yemen PD Rep.	Asia	not identified	897	not identified
Kyrgysztan	Europe	7,000,000,000	39,400	0.000563
Moldova	Europe	10,000,000,000	18,000	0.000180
Russian Federation	Europe	480,000,000,000	20,297,881	0.004229
Tajikistan	Europe	6,000,000,000	93,425	0.001557
Turkey	Europe	91,700,000,000	1,655	0.000002
Italy	Europe	971,000,000,000	700,000	0.000072
Australia	Oceania	291,000,000,000	176,646	0.000061
Fiji	Oceania	1,300,000,000	25,794	0.001984
Papua New Guinea	Oceania	3,400,000,000	41,640	0.001225
Vanuatu	Oceania	200,000,000	10,732	0.005366
Western Samoa**	Oceania	159,000,000	39,700	0.024969

* GNP 1990; Source: UNDP, Human Development Report 1993.
** GNP 1993; Source: OECD, DAC Report 1995.

Economic data is only available for a limited number of disaster-affected countries. Predictably, the larger recorded economic costs come from the more developed countries with their more costly infrastructure. However, for small states like Vanuatu and Western Samoa, disasters can absorb a disproportionalty high percentage of gross national product. Source: CRED

Table 13 Refugees and asylum seekers by country of origin

	1990	1991	1992	1993	1994	1995	1996
AFRICA	**5,414,900**	**5,321,500**	**5,730,600**	**5,812,400**	**5,857,650**	**5,191,200**	**3,526,000**
Angola	435,700	443,200	404,200	335,000	344,000	313,000	190,000
Burundi	186,200	208,500	184,000	780,000	330,000	290,000	285,000
Chad	34,400	34,800	24,000	33,400	29,000	16,000	16,000
Djibouti	—	—	—	7,000	10,000	10,000	10,000
Eritrea	*	*	*	373,000	384,500	342,500	340,000
Ethiopia	1,066,300	752,400	834,800	232,200	190,750	110,700	75,000
Liberia	729,800	661,700	599,200	701,000	784,000	725,000	730,000
Mali	21,400	53,000	81,000	87,000	115,000	90,000	85,000
Mauritania	60,100	66,000	65,000	79,000	75,000	80,000	80,000
Mozambique	1,427,500	1,483,500	1,725,000	1,332,000	325,000	970,000	—
Niger	3,500	500	5,000	6,000	20,000	20,000	20,000
Rwanda	203,900	203,900	201,500	275,000	1,715,000	1,545,000	250,000
Senegal	24,400	27,600	15,000	18,000	17,000	17,000	15,000
Sierra Leone	—	181,000	200,000	260,000	260,000	363,000	345,000
Somalia	454,600	717,600	864,800	491,200	457,400	480,300	440,000
South Africa	40,000	23,700	11,100	10,600	—	—	—
Sudan	499,100	202,500	263,000	373,000	510,000	448,100	430,000
Togo	—	15,000	6,000	240,000	140,000	95,000	30,000
Uganda	12,300	14,900	15,100	20,000	15,000	10,000	15,000
Western Sahara	165,000	165,000	165,000	80,000	80,000	80,000	80,000
Zaire	50,700	66,700	66,900	79,000	56,000	58,600	90,000
EAST ASIA AND PACIFIC	**698,900**	**811,450**	**502,000**	**797,400**	**690,050**	**640,950**	**613,100**
Burma	50,800	112,000	86,700	289,500	203,300	160,400	136,500
Cambodia	344,500	392,700	148,600	35,500	30,250	26,300	34,400
China (Tibet)	114,000	114,000	128,000	133,000	139,000	141,000	141.000
Indonesia	—	6,900	5,500	9,400	9,700	9,500	9,500
Laos	67,400	63,200	43,300	26,500	12,900	8,900	3,500
Viet Nam	122,200	122,650	89,900	303,500	294,900	294,850	288,200
SOUTH AND CENTRAL ASIA	**6,330,100**	**6,900,800**	**4,715,400**	**3,899,050**	**3,319,200**	**2,809,400**	**3,112,400**
Afghanistan	6,027,100	6,600,800	4,286,000	3,429,800	2,835,300	2,328,400	2,552,800
Bangladesh	75,000	65,000	50,000	53,500	48,300	48,000	48,000
Bhutan	—	25,000	95,400	105,100	116,600	118,600	119,000
Sri Lanka	228,000	210,000	181,000	106,650	104,000	96,000	101,000
Tajikistan	*	*	52,000	153,000	165,000	170,400	240,000
Uzbekistan	*	*	51,000	51,000	50,000	48,000	51,600

Source: US Committee for Refugees

MIDDLE EAST	**3,554,400**	**2,792,500**	**2,849,300**	**2,975,000**	**3,826,950**	**3,958,500**	**4,374,900**
Iran	211,100	50,000	65,400	39,000	54,250	49,500	45,800
Iraq	529,700	217,500	125,900	134,700	635,900	622,900	639,600
Kuwait	385,500	—	—	—	—	—	—
Palestine	2,428,100	2,525,000	2,658,000	2,801,300	3,136,800	3,266,100	3,689,600
EUROPE	**0**	**120,000**	**2,529,800**	**1,952,650**	**1,775,800**	**1,805,600**	**1,895,800**
Armenia	*	*	202,000	200,000	229,000	185,000	198,800
Azerbaijan	*	*	350,000	290,000	374,000	390,000	325,000
Bosnia and Herzegovina#	*	*	n.a.	n.a.	863,300	905,500	944,000
Croatia #	*	*	n.a.	n.a.	136,900	200,000	300,000
Georgia	*	*	130,000	143,000	106,800	105,000	108,000
Moldova	*	*	80'000	—	—	—	—
Turkey	—	—	—	—	13,000	15,000	15,000
Yugoslavia #	—	120,000	1,767,800	1,319,650	52,800	5,100	5,000
AMERICAS AND THE CARIBBEAN	**152,400**	**118,250**	**104,250**	**97,500**	**120,550**	**68,400**	**81,700**
Colombia	3,000	4,000	—	450	100	200	700
Cuba	2,900	1,400	1,650	1,400	30,600	4,000	4,000
El Salvador	37,200	24,200	22,800	21,900	16,200	12,400	12,000
Guatemala	57,400	46,700	45,750	49,200	45,050	34,150	34,000
Haiti	—	6,950	1,600	1,500	5,850	1,500	15,000
Nicaragua	41,900	25,400	30,850	23,050	22,750	16,150	16,000
Suriname	10,000	9,600	1,600	—	—	—	—
WORLD TOTAL	**16,150,700**	**16,064,500**	**16,431,350**	**15,534,000**	**15,590,200**	**14,479,850**	**13,603,900**

Notes: — indicates zero or near zero; * country did not exist as of reporting date; n.a. not available, or reported estimates unreliable; # for 1992-93, refugees from Croatia and Bosnia included in Yugoslavia total, for 1994-95, Yugoslavia total includes only refugees from Serbia and Montenegro.

> Global refugee figures continue to show last year's trend with a slight decease in the year-on-year figure. In Africa, figures are down by nearly two million reflecting the return home of people to Rwanda and Angola. The figures also clearly show the intractable nature of many refugee problems with the numbers of refugees and asylum seekers failing to return home remaining constant over the years for many countries.Source: US Committee for Refugees

TABLE 14 Refugees and asylum seekers by host country

	1990	1991	1992	1993	1994	1995	1996
AFRICA	**5,442,450**	**5,339,950**	**5,697,650**	**5,824,700**	**5,879,700**	**5,222,300**	**3,657,200**
Algeria	189,400	204,000	210,000	121,000	130,000	120,000	100,000
Angola	11,900	10,400	9,000	11,000	11,000	10,900	9,300
Benin	800	15,100	4,300	120,000	50,000	25,000	15,000
Botswana	1,000	1,400	500	500	—	—	—
Burkina Faso	300	400	6,300	6,000	30,000	21,000	10,000
Burundi	90,700	107,000	107,350	110,000	165,000	140,000	12,000
Cameroon	6,900	6,900	1,500	2,500	2,000	2,000	2,000
Cent. African Rep.	6,300	9,000	18,000	41,000	42,000	34,000	36,400
Congo	3,400	3,400	9,400	13,000	16,000	15,000	16,000
Côte d'Ivoire	270,500	240,400	195,500	250,000	320,000	290,000	300,000
Djibouti	67,400	120,000	96,000	60,000	60,000	25,000	20,000
Egypt	37,800	7,750	10,650	11,000	10,700	10,400	46,000
Ethiopia	783,000	534,000	416,000	156,000	250,000	308,000	320,000
Gabon	800	800	200	200	—	1,000	1,000
Gambia	800	1,500	3,300	2,000	1,000	5,000	5,000
Ghana	8,000	6,150	12,100	133,000	110,000	85,000	90,000
Guinea	325,000	566,000	485,000	570,000	580,000	640,000	620,000
Guinea-Bissau	1,600	4,600	12,000	16,000	16,000	15,000	15,000
Kenya	14,400	107,150	422,900	332,000	257,000	225,000	185,000
Lesotho	1,000	300	200	100	—	—	—
Liberia	—	12,000	100,000	110,000	100,000	120,000	100,000
Libya	—	—	—	—	—	28,100	27,200
Malawi	909,000	950,000	1,070,000	700,000	70,000	2,000	—
Mali	10,600	13,500	10,000	13,000	15,000	15,000	15,000
Mauritania	22,000	40,000	40,000	46,000	55,000	35,000	25,000
Mozambique	700	500	250	—	—	—	—
Namibia	25,000	30,200	150	5,000	1,000	1,000	1,000
Niger	800	1,400	3,600	3,000	3,000	17,000	15,000
Nigeria	5,300	4,600	2,900	4,400	5,000	8,000	8,000
Rwanda	21,500	32,500	24,500	370,000	—	—	20,000
Senegal	55,300	53,100	55,100	66,000	60,000	68,000	68,000
Sierra Leone	125,000	17,200	7,600	15,000	20,000	15,000	15,000
Somalia	358,500	35,000	10,000		—	—	—
South Africa	201,000	201,000	250,000	300,000	200,000	90,000	20,000
Sudan	726,500	717,200	750,500	633,000	550,000	450,000	415,000
Swaziland	47,200	47,200	52,000	57,000	—	—	—
Tanzania	266,200	251,100	257,800	479,500	752,000	703,000	325,000
Togo	—	450	350	—	5,000	10,000	10,000

Source: US Committee for Refugees

Tunisia	—	—	—	—	—	500	300
Uganda	156,000	165,450	179,600	257,000	323,000	230,000	225,000
Zaire	370,900	482,300	442,400	452,000	1,527,000	1,332,000	465,000
Zambia	133,950	140,500	155,700	158,500	123,000	125,400	100,000
Zimbabwe	186,000	198,500	265,000	200,000	20,000	—	—
EAST ASIA AND PACIFIC	**592,100**	**688,500**	**398,600**	**467,600**	**444,100**	**452,850**	**454,750**
Australia	n.a.	23,000	24,000	2,950	5,300	7,500	12,900
China	5,000	14,200	12,500	296,900	297,100	294,100	294,200
Hong Kong	52,000	60,000	45,300	3,550	1,900	1,900	1,300
Indonesia	20,500	18,700	15,600	2,400	250	—	—
Japan	800	900	700	950	7,350	9,900	400
Korea	200	200	150	—	—	—	—
Macau	200	100	—	—	—	—	—
Malaysia	14,600	12,700	16,700	8,150	6,100	5,300	5,000
Papua New Guinea	8,000	6,700	3,800	7,700	9,700	9,500	9,500
Philippines	19,600	18,000	5,600	1,700	250	450	50
Singapore	150	150	100	—	—	—	—
Solomon Islands	—	—	—	—	3,000	1,000	1,000
Taiwan	150	150	150	—	—	—	—
Thailand	454,200	512,700	255,000	108,300	83,050	98,200	96,000
Viet Nam	16,700	21,000	19,000	35,000	30,100	25,0000	34,400
EUROPE	**627,400**	**578,400**	**3,210,400**	**2,542,100**	**2,421,500**	**2,520,700**	**2,409,350**
Armenia	*	*	300,000	290,000	295,800	304,000	150,000
Austria	22,800	27,300	82,100	77,700	59,000	55,900	77,000
Azerbaijan	*	*	246,000	251,000	279,000	238,000	248,000
Belarus	*	*	3,700	10,400	18,800	7,000	36,000
Belgium	13,000	15,200	19,100	32,900	19,400	16,400	24,900
Bulgaria	—	—	—	—	900	500	550
Croatia	*	*	420,000	280,000	188,000	189,500	140,000
Czechoslovakia	1,600	2,800	2,200	*	*	*	*
Czech Republic	*	*	*	6,300	4,700	2,400	1,550
Denmark	5,500	4,600	13,900	23,300	24,750	9,600	4,400
Estonia	*	*	—	—	100	—	—
Finland	2,700	2,100	3,500	3,700	850	750	600
France	56,000	46,800	29,400	30,900	32,600	30,000	28,000
Germany	193,100	256,100	536,000	529,100	430,000	442,700	370,000
Greece	6,200	2,700	1,900	800	1,300	3,200	1,600
Hungary	18,300	5,200	40,000	10,000	11,200	9,100	5,000
Ireland	—	—	—	—	—	—	1,000
Italy	4,800	31,400	19,100	33,550	31,800	60,700	60,600

Source: US Committee for Refugees

Latvia	*	*	—	—	150	150	—
Lithuania	*	*	—	—	—	400	1,500
Macedonia	*	*	32,700	12,100	8,200	7,000	5,100
Netherlands	21,200	21,600	24,600	35,400	52,600	39,300	31,000
Norway	3,900	4,600	5,700	14,200	11,600	11,200	12,500
Poland	—	2,500	1,500	600	500	800	1,300
Portugal	100	200	—	2,250	600	350	150
Romania	—	500	—	1,000	600	1,300	200
Russian Federation	*	*	460,000	347,500	451,000	500,000	522,000
Slovak Republic	*	*	*	1,900	2,000	1,600	1,700
Slovenia	*	*	68,900	38,000	29,000	24,000	10,300
Spain	6,800	8,100	12,700	14,000	14,500	4,300	4,900
Sweden	28,900	27,300	88,400	58,800	61,000	12,300	7,500
Switzerland	37,000	41,600	81,700	27,000	23,900	29,000	34,400
Turkey	178,000	31,500	31,700	24,600	30,650	21,150	13,000
Ukraine	*	*	40,000	—	5,000	6,000	8,000
United Kingdom	25,000	44,700	24,600	28,100	32,000	44,000	56,600
Yugoslavia #	2,500	1,600	621,000	357,000	300,000	450,000	550,000
AMERICAS AND CARIBBEAN	**229,050**	**218,800**	**249,000**	**272,450**	**297,300**	**256,400**	**564,300**
Argentina	1,800	1,800	—	—	—	—	—
Bahamas	—	—	—	—	—	200	200
Belize	6,200	12,000	8,700	8,900	8,800	8,650	8,600
Bolivia	100	100	—	600	600	600	600
Brazil	200	200	200	1,000	2,000	2,000	2,000
Canada	36,600	30,500	37,700	20,500	22,000	24,900	30,800
Chile	—	—	—	100	200	300	300
Colombia	700	700	400	400	400	400	400
Costa Rica	26,900	24,300	34,350	24,800	24,600	20,500	20,000
Cuba	3,000	1,100	1,100	—	—	1,800	1,800
Dominican Republic	—	—	—	1,300	1,350	900	600
Ecuador	3,750	4,200	200	100	100	100	100
El Salvador	600	250	250	150	150	150	150
French Guiana	10,000	9,600	1,600	—	—	—	—
Guatemala	6,700	8,300	4,900	4,700	4,700	2,500	2,500
Honduras	2,700	2,050	150	100	100	50	50
Mexico	53,000	48,500	47,300	52,000	47,700	38,500	38,000
Nicaragua	500	2,800	5,850	4,750	300	450	400
Panama	1,200	1,300	850	950	900	800	700
Peru	600	600	400	400	700	700	700
United States	73,600	68,800	103,700	150,400	181,700	152,200	455,700
Venezuela	900	1,700	1,350	1,300	1,000	700	700

Source: US Committee for Refugees

MIDDLE EAST	5,698,600	5,770,200	5,586,850	4,923,800	5,447,750	5,499,100	5,799,300
Bahrain	7,500	—	—	—	—	—	—
Gaza Strip	496,300	528,700	560,200	603,000	644,000	683,600	716,900
Iran	2,860,000	3,150,000	2,781,800	1,995,000	2,220,000	2,075,500	2,020,000
Iraq	60,000	48,000	64,600	39,500	120,500	115,200	114,400
Jordan	929,100	960,200	1,010,850	1,073,600	1,232,150	1,294,800	1,362,500
Kuwait	—	—	—	—	25,000	55,000	42,000
Lebanon	306,400	314,200	322,900	329,000	338,200	348,300	355,000
Oman	3,000	—	—	—	—	—	—
Saudi Arabia	300,000	34,000	27,400	25,000	17,000	13,200	257,800
Syria	280,700	293,900	307,500	319,200	332,900	342,300	324,200
United Arab Emirates	40,000	—	—	—	150	400	400
West Bank	414,300	430,100	459,100	479,000	504,000	517,400	532,400
Yemen	1,300	11,100	52,500	60,500	13,850	53,400	50,500
SOUTH AND CENTRAL ASIA	4,098,600	4,050,750	2,341,700	2,151,400	1,776,450	1,386,300	1,642,500
Afghanistan	—	—	52,000	35,000	20,000	18,400	18,900
Bangladesh	—	30,150	245,300	199,000	116,200	55,000	30,500
India	415,800	402,600	378,000	325,600	327,850	319,200	324,200
Kazakhstan	*	*	—	6,500	300	6,500	14,000
Kyrgyzstan	*	*	—	3,500	350	7,600	16,700
Nepal	14,000	24,000	89,400	99,100	104,600	106,600	109,800
Pakistan	3,668,800	3,594,000	1,577,000	1,482,300	1,202,650	867,500	1,102,600
Tajikistan	*	*	—	400	2,500	2,500	1,000
Turkmenistan	*	*	—	—	—	—	22,000
Uzbekistan	*	*	—	—	2,000	3,000	2,800
WORLD TOTAL	16,688,200	16,646,60	17,484,200	16,182,05	16,266,800	15,337,650	14,527,400

Notes: — indicates zero or near zero; * country did not exist as of reporting date; n.a. not available, or reported estimates unreliable; # for 1992-93, refugees from Croatia and Bosnia included in Yugoslavia total, for 1994-95, Yugoslavia total includes only refugees from Serbia and Montenegro.

> *Many countries both generate and host refugees. Ethiopia generated 75,000 refugees in 1996 and hosted 320,000. Iraq hosted 114,000 and generated nearly 640,000. Malawi, after over a decade of hosting one of the world's largest refugee populations, shows a zero "host" figure for the first time while other countries, particularly those of the former Soviet Union Lithuania and Turkmenistan, for example become hosts to refugees for the first time. Source: US Committee for Refugees*

TABLE 15 Significant populations of internally-displaced people

	1990	1991	1992	1993	1994	1995	1996
AFRICA	**13,504,000**	**14,222,000**	**17,395,000**	**16,890,000**	**15,730,000**	**10,185,000**	**9,380,000**
Algeria	—	—	—	—	—	—	10,000
Angola	704,000	827,000	900,000	2,000,000	2,000,000	1,500,000	1,200,000
Burundi	—	—	—	500,000	400,000	300,000	400,000
Djibouti	—	—	—	140,000	50,000	—	—
Eritrea	*	*	*	200,000	—	—	—
Ethiopia	1,000,000	1,000,000	600,000	500,000	400,000	—	—
Ghana	—	—	—	—	20,000	150,000	100,000
Kenya	—	—	45,000	300,000	210,000	210,000	150,000
Liberia	500,000	500,000	600,000	1,000,000	1,100,000	1,000,000	1,000,000
Mozambique	2,000,000	2,000,000	3,500,000	2,000,000	500,000	500,000	300,000
Rwanda	—	100,000	350,000	300,000	1,200,000	500,000	—
Sierra Leone	—	145,000	200,000	400,000	700,000	1,000,000	1,000,000
Somalia	400,000	500,000	2,000,000	700,000	500,000	300,000	250,000
South Africa	4,100,000	4,100,000	4,100,000	4,000,000	4,000,000	500,000	500,000
Sudan	4,500,000	4,750,000	5,000,000	4,000,000	4,000,000	4,000,000	4,000,000
Togo	—	—	—	150,000	100,000	—	—
Uganda	300,000	300,000		—	—	—	70,000
Zaire	—	—	100,000	700,000	550,000	225,000	400,000
AMERICAS AND CARIBBEAN	**1,126,000**	**1,221,000**	**1,354,000**	**1,400,000**	**1,400,000**	**1,280,000**	**1,280,000**
Colombia	50,000	150,000	300,000	300,000	600,000	600,000	600,000
El Salvador	400,000	150,000	154,000	—	—	—	—
Guatemala	100,000	150,000	150,000	200,000	200,000	200,000	200,000
Haiti	—	200,000	250,000	300,000	—	—	—
Honduras	22,000	7,000	—	—	—	—	—
Nicaragua	354,000	354,000	—	—	—	—	—
Panama	—	10,000	—	—	—	—	—
Peru	200,000	200,000	500,000	600,000	600,000	480,000	480,000
SOUTH AND CENTRAL ASIA	**3,085,000**	**2,685,000**	**1,810,000**	**880,000**	**1,775,000**	**1,600,000**	**1,770,000**
Afghanistan	2,000,000	2,000,000	530,000	n.a.	1,000,000	500,000	500,000
India	85,000	85,000	280,000	250,000	250,000	250,000	250,000
Sri Lanka	1,000,000	600,000	600,000	600,000	525,000	850,000	1,000,000
Tajikistan	*	*	400,000	30,000	—	—	20,000

Source: US Committee for Refugees

EUROPE	1,048,000	1,755,000	1,626,000	2,765,000	5,195,000	5,080,000	4,740,000
Armenia	*	*	*	—	—	75,000	50,000
Azerbaijan	*	*	216,000	600,000	630,000	670,000	550,000
Bosnia and Herzegovina	*	*	740,000	1,300,000	1,300,000	1,300,000	1,110,000
Croatia	*	*	340,000	350,000	290,000	240,000	110,000
Cyprus	268,000	268,000	265,000	265,000	265,000	265,000	265,000
Georgia	*	*	15,000	250,000	260,000	280,000	285,000
Moldova	*	*	20,000	—	—	—	—
Russian Federation	*	*	—	n.a.	450,000	250,000	380,000
Soviet Union	750,000	900,000	*	*	*	*	*
Turkey	30,000	30,000	30,000	n.a.	2,000,000	2,000,000	2,000,000
Yugoslavia #	—	557,000	—	—	—	—	—
MIDDLE EAST	1,300,000	1,450,000	800,000	1,960,000	1,710,000	1,700,000	1,475,000
Iran	—	—	—	260,000	—	—	—
Iraq	500,000	700,000	400,000	1,000,000	1,000,000	1,000,000	900,000
Lebanon	800,000	750,000	400,000	700,000	600,000	400,000	450,000
Yemen	—	—	—	—	110,000	300,000	125,000
EAST ASIA AND PACIFIC	340,000	680,000	699,000	595,000	613,000	555,000	555,000
Burma	200,000	500,000	500,000	500,000	500,000	500,000	500,000
Cambodia	140,000	180,000	199,000	95,000	113,000	55,000	55,000
WORLD TOTAL	20,403,000	22,013,000	23,684,000	24,490,000	26,423,000	20,400,000	19,195,000

Notes: — indicates zero or near zero; * country did not exist as of reporting date; n.a. not available, or reported estimates unreliable;
for 1992-93, refugees from Croatia and Bosnia included in Yugoslavia total, for 1994-95, Yugoslavia total includes only refugees from Serbia and Montenegro.

Numbers of internally-displaced people have dropped from 10.2 million in 1995 to 9.4 million in 1996. In Africa the half-million people recorded as internally-displaced in Rwanda have now returned home, while the Zaire figure has gone up to 400,000, reflecting the growing civil unrest there. In Europe, the figure for Bosnia and Herzegovina remains high at 1.1 million but shows a slight fall from 1995's figures as people struggle to restore normality after the war. The figure of two million displaced in Turkey reflects the continuing, much less-publicized, internal strife of that country. Source: US Committee for Refugees

Table 16 Number of major armed conflicts by region per year over seven years (1990-1996)

	1990	1991	1992	1993	1994	1995	1996
Europe / Total	1	2	4	6	5	3	2
War	0	1	2	4	1	2	1
Intermediate	1	1	2	2	4	1	1
Middle East / Total	5	7	5	6	6	6	6
War	1	3	1	1	2	1	1
Intermediate	4	4	4	5	4	5	5
Asia / Total	15	12	13	11	11	12	11
War	6	7	7	4	2	2	2
Intermediate	9	5	6	7	9	10	9
Africa / Total	11	11	7	7	7	6	6
War	9	9	7	4	2	2	3
Intermediate	2	2	0	3	5	4	3
Americas / Total	4	4	3	3	3	3	3
War	3	1	3	2	0	0	0
Intermediate	1	3	0	1	3	3	3

An armed conflict is defined as "major" when at least 1,000 battle-related deaths have occurred since the beginning of conflict. Major armed conflicts are divided into two categories: war (more than 1,000 battle-related deaths during the year in question); and, intermediate (less than 1,000 battle-related deaths in a given year).

All regions of the world have witnessed at least one major armed conflict during the 1990s. Not unexpectedly, Asia and Africa — the largest regions in population and territory respectively — consistently showed the greatest number of wars, although the number has declined in both regions over the period in question. In South-East Asia and in Southern Africa, especially, there are fewer conflicts and of a lesser intensity.

In Europe, after reaching a peak in 1993 and 1994, the number of major armed conflicts continued to fall from a peak of six in 1993 to two in 1996 (Chechnya and Northern Ireland). By the end of the year, however, the fighting in Chechnya had ceased.

In the Americas, a reduction in the intensity of conflicts meant a considerable shift from wars to intermediate armed conflicts, although the total numbers declined only slightly. Unlike the other regions, the Middle East showed no significant decline in either the number of or the intensity of the armed conflicts recorded for the period. Source: DPCR

TABLE 17 Number of battle-related deaths in major armed conflicts per region per year over seven years (1990-1996)

	1990	1991	1992	1993	1994	1995	1996
Europe	74	6,000 -10,000	11,200 - 21,400	14,200 - 42,000	>1,500	11,000 - 43,000	>2,000
Middle East	>3,500	>16,000	3,300 - 4,500	3,000 - 4,000	4,800 - 12,000	4,250 - 5,500	>2,500
Asia	>15,000	>16,000	14,000 - 60,000	23,500 - 35,000	6,300 - 15,000	>7,500	>5,500
Africa	>33,500	>37,000	14,000 - 40,000	>25,500	25,000 - 35,000	>5,500	>6,000
Americas	6,000 - 7,500	3,200 - 6,200	>5,400	<3,400	<1,400	<1,700	450 - 1200

The highest figures for battle-related casualties are in Africa and in Asia. The intensity of wars in these two regions varied at different times during the period in question.

In Africa, for example, the wars in the Horn of Africa were particularly severe in the first years of the 1990s; later, the war in Algeria has resulted in high casualties; and 1996 saw an escalation of the old conflict in Sudan. The conflagration over Eastern Zaire, which received world attention in October and November 1996, saw limited fighting. The advances of the opposition group AFDL (Alliance des forces démocratiques pour la libération du Congo-Kinshasa) did not involve heavy battles. It resulted, however, in the displacement and ultimate return of hundreds of thousands of Rwandese refugees. The attacks in Burundi, involving the Tutsi-dominated army and Hutu-led militants were often directed against civilians; there were few battles in a strict sense. Zaire and Burundi are therefore not included in the tables. The Great Lakes region of Central Africa was of major concern, however, to African leaders as well as the outside world.

In Asia, most of the battle-related deaths stem from the wars in Afghanistan, India, Sri Lanka and Tajikistan. The peak in 1992-1993 was caused by the war in Tajikistan.

The record of battle-related deaths for Europe shows two peaks, one connected with the war in former Yugoslavia (mostly Bosnia and Herzegovina, particularly in 1992-1993) and one with the war in Chechnya, 1995.

For the Middle East there is a peak in 1991 (i.e. the Gulf War) and in 1994 (the war in Yemen).

For Central and South America, the steady reduction of battle-related deaths continued throughout the period. In sum, there has been a general decline in battle-related deaths in all regions in 1995-1996.
Source: DPCR

Table 18 Number of major armed conflicts by type of incompatibility per region per year from 1990 to 1996

	1990		1991		1992		1993		1994		1995		1996	
	G	T	G	T	G	T	G	T	G	T	G	T	G	T
Europe	—	1	—	2	—	4	—	6	—	5	—	3	—	2
Middle East	1	4	2	5	2	3	2	4	2	4	2	4	2	4
Asia	5	10	3	8	4	9	4	7	4	7	4	8	4	7
Africa	8	3	8	3	6	1	6	1	6	1	5	1	5	1
Americas	4	—	4	—	3	—	3	—	3	—	3	—	3	—

Major armed conflicts can be divided into two categories: Government (G) — a war whose aim is to change the type of political system, the central government or its composition, or Territory (T) — a conflict concerning control of territory (interstate conflict), secession or autonomy.

Although conflicts to gain control over government or over a specified territory are almost equally common, globally speaking, there are some regional differences.

The major armed conflicts in Europe have all concerned territory: the wars in Northern Ireland, the former Yugoslavia and in the successor states of the Soviet Union dealt with groups trying to gain control over territory. In some cases, the desire for independence led to conflict, for example, Nagorno-Karabach in Azerbaijan, and Chechnya in the Russian Federation. In others, such as the protracted war in Bosnia and Herzegovina, complications were many as proponents of opposing state-building projects clashed. Compared to other parts of the world, Europe stands out as a region marked with such nationalist struggles. In the Americas, the pattern was the opposite to that of Europe: all conflicts concerned the control over government. The issue was often defined along a left-right continuum.

In the Middle East and Asia, territorial disputes dominated; for example, the Palestinian and Kurdish conflicts, and the fighting in Punjab, Kashmir, East Timor and Mindanao. In both regions, wars were fought for control of the government, e.g., Iran, Iraq, Yemen and the devastating conflicts in Afghanistan and Tajikistan.

In Africa, too, the cause of most conflicts was governmental control, although the mobilization often followed ethnic lines. In Somalia, Rwanda and Liberia, for instance, ethnically-based groups fought to control central power, rather than fighting for the dismemberment of the state. In some of these cases, the weakness of the state in a society divided between ethnic groups or strong clans, might make it more tempting to try to gain control over the state as a whole, rather than for only a part of the state. However, the very same divisions of the society may make victory difficult to achieve and the result is protracted conflicts. Source: DPCR

TABLE 19 Non-food emergency and distress relief, grant disbursements in millions US$ from 1985 to 1995

DONOR	1985	1986	1987	1988	1989	1990	1991	1992	1993	1994	1995
Australia	8.81	5.90	17.31	7.94	6.78	12.23	13.23	29.56	26.56	25.49	35.80
Austria	3.36	3.03	5.92	14.07	22.68	43.96	93.87	145.83	123.45	127.04	114.72
Belgium	1.92	2.12	1.21	1.75	1.59	4.59	5.71	13.18	19.05	14.02	15.75
Canada	54.66	25.88	26.44	55.86	29.69	45.76	85.14	78.86	273.96	228.45	164.72
Denmark	—	—	—	—	—	108.27	52.83	104.85	77.14	78.62	71.38
Finland	5.03	9.81	26.21	19.74	31.96	70.54	102.24	61.55	21.61	27.48	22.64
France	—	—	—	—	—	—	—	25.88	125.08	122.223	138.43
Germany	18.01	22.44	28.16	35.82	30.68	45.21	415.31	680.32	549.52	392.53	438.71
Ireland	1.32	1.08	1.06	1.22	1.33	2.09	2.89	2.10	5.15	8.53	8.34
Italy	90.66	188.44	124.88	145.21	84.15	104.06	456.33	137.39	341.69	105.40	87.89
Japan	6.94	1.87	2.22	8.90	19.61	26.46	20.48	14.93	40.37	31.08	60.08
Luxembourg	—	—	0.40	1.50	2.00	3.80	10.30	7.21	8.49	5.09	7.03
Netherlands	22.92	25.37	28.96	33.99	24.40	63.58	109.74	197.45	303.29	302.37	350.42
New Zealand	0.91	0.20	1.24	0.61	—	3.85	1.51	5.12	4.96	2.68	1.84
Norway	21.03	22.44	20.37	41.74	50.38	88.62	77.60	86.48	113.21	180.75	183.78
Portugal	—	—	—	—	—	—	0.11	0.11	8.35	3.70	3.52
Spain	—	—	—	—	1.20	5.00	8.42	6.43	7.74	5.04	19.53
Sweden	78.90	99.34	133.29	110.28	214.50	124.49	181.65	342.56	277.28	334.17	269.75
Switzerland	19.14	25.90	64.53	40.87	46.49	46.75	67.78	68.61	66.85	80.98	97.20
United Kingdom	43.03	27.21	21.49	32.36	31.72	37.95	116.48	56.83	187.27	260.52	181.76
United States	225.00	193.00	183.00	170.00	210.00	221.00	596.00	521.00	669.00	1,132.00	789.00
TOTAL	**601.64**	**654.03**	**686.69**	**721.86**	**809.16**	**1,058.21**	**2,417.62**	**2,586.25**	**3,250.02**	**3,468.17**	**3,062.29**
Proportion for refugees	573.01	627.42	130.30	177.47	225.76	348.21	1,052.41	1,713.30	1,976.28	1,976.28	1,799.49

> *Figures for the economic inputs to humanitarian response both cash disbursements and food aid take longer to generate from the statistical systems of the intergovernmental organizations than data on people affected, and hence are only reported here up to 1995; 1996's figures were still very provisional as the World Disasters Report 1997 went to press. However, it is clear that the run-away spending in international humanitarian action is being checked. The global figure for 1995 is significantly down on 1994's record high, and provisional figures for 1996 confirm this trend (see chapter 5).*
> *Source: OECD/DAC.*

Table 20 Breakdown of food-aid deliveries by category per year from 1987 to 1995 in thousand tonnes – cereals in grain equivalent

	1987	1988	1989	1990	1991	1992	1993	1994	1995
Emergency	1,987	3,281	2,389	2,767	3,540	4,991	4,202	4,208	3,451
Project	3,938	4.057	2,872	2,837	2,980	2,578	2,498	2,779	2,408
Programme	8,479	7,510	6,473	8,038	6,650	7,663	10,170	5,730	4,046
World total	14,404	14,848	11,734	13,642	13,170	15,232	16,870	12,717	9,905

Figures for food-aid deliveries are well down, reflecting the reduced grain surpluses generated, as world trading agreements leads to a sharper relationship between the amount of grain grown and the amount the world market can purchase. As a consequence of this, an increasing proportion of food aid is focused on emergency response rather than long-term development.
Source: WFP

Chapter 10 Data sources

CRED
Catholic University of Louvain
School of Public Health
Clos Chapelle aux Champs 30-34
1200 Brussels
Belgium
Tel.: (32)(2) 764 3327
Fax : (32)(2) 764 3328
E-mail: misson@epid.ucl.ac.be

US Committee for Refugees
1717 Massachusetts Ave NW
Suite 701
Washington DC 20036
USA
Tel.: (1)(202) 347 3507
Fax: (1)(202) 347 3418
E-mail: uscr@irsa-uscr.org

Department of Peace and Conflict
Research
Uppsala University
Box 514
S-751 20 Uppsala
Sweden
Tel.: (46)(18) 182 352
Fax: (46(18) 695 102
E-mail:
Peter.Wallensteen@pcr.uu.se

DAC/OECD
2, rue André-Pascal
75775 Paris Cedex 16
France
Tel.: (33)(1) 4524 8980
Fax: (33)(1) 4524 1650

Food Aid Information Group
World Food Programme
Via Cristoforo Colombo 426
00145 Rome
Italy
Tel.: (39(6) 5228 2796
Fax: (39(6) 5228 2451
E-mail: simon@wfp.org

Aid agencies should be about ethics and putting values into action, not just shifting food aid. In a world where it sometimes seems the "bottom line" is all, how can value-based organizations ensure their actions live up to their rhetoric?
Bosnia, 1996.

Chapter

11 From codes of conduct to standards of performance

The increasing scale, complexity, speed and cost of emergencies mean that humanitarian agencies must confront questions of the quality of their work and issues of success or failure, even whether what they are doing is right or wrong. Humanitarian aid, like the nation state and the laws of war, is a device to suit a time. Its form and applicability are not sacrosanct. Is humanitarian aid the best way to address crises today?

It is important to define what humanitarianism is not. Humanitarian does not mean caring, alleviating poverty or providing relief, although it may include elements of all these. "Humanity" is about respect for the whole human being, for everything which constitutes a person's being, so humanitarianism is not limited to relief aid. Instead, humanitarianism is a way of acting: carrying out actions which are, and are perceived to be, impartial, neutral and, by extension, independent from political, religious or other extraneous bias.

This was at the core of an agreement in 1859 between Swiss pacifist Henry Dunant and French statesman Napoleon III, which is still having ramifications for how humanitarians should act today. Dunant convinced Napoleon, victorious in the battle of Solferino, of the moral correctness of helping the wounded on the battlefield, regardless of nationality.

Napoleon turned goodwill into an issue of rights and justice by allowing assistance to be delivered under the protection of an official proclamation. It is from this beginning that The Hague and Geneva Conventions, and many of today's declarations on humanitarian issues, stem.

Relief aid is not necessarily humanitarian. It is not humanitarian when armies relieve cities under siege, providing food, water and medical care in an emergency. It is not humanitarian when development agencies seek to address the root causes of poverty, tackling issues of land tenure, access to education or freedom of speech, in an impartial fashion but still sitting in judgement as to who is right and who is wrong.

By adhering to the principles of impartiality, neutrality and independence, humanitarian organizations limit themselves to addressing the effects of crisis – not directly the causes – and to use only tools that conform to those principles. Limits confine, but they also safeguard both the rights of individuals caught up in crisis, regardless of their past and affiliations, and the ability of the agency to continue to deliver its assistance and protection in the future.

> *In most disaster situations, the actions of one NGO very quickly reflect on the credibility of all others.*

Two main ethical approaches are found in aid, both derived from the moral impulse to act to help others. Many individuals do what they do because they care. Seeing suffering, they are impelled to act. The moral basis of the actions is clear and can be found in Christianity, Islam, Buddhism and many other beliefs.

One approach holds that there is an imperative or duty to act and morality rests in the act. Thus – in the language of the Red Cross and Red Crescent – the duty of a humanitarian is "to alleviate suffering regardless of race, creed or political persuasion". Humanitarians are driven by the immediacy and the rightness of the act, not its potential consequences.

This tradition underlies the Geneva Conventions, from which much present-day humanitarian action is derived, making it an issue of justice involving non-negotiable rights. In this tradition, humanitarian agencies forfeit the right to be concerned with the cause of suffering so that they may continue to alleviate suffering wherever and whenever it occurs.

This does not free humanitarians from considering the consequences of their actions. They may not design them to have developmental or political effects, but humanitarians have to be sure that their actions, at the very least, do no harm to people's long-term prospects. Neither does it free humanitarians from being concerned with justice, although the concern has to be to support the process of justice, not passing judgement. Humanitarians do not decide who is guilty or innocent.

The other ethical approach sees the civil, political and economic consequences of actions – not the actions themselves – as the test of moral good or bad. The good of the act has to be weighed against the greater good for the family, community, state or wider world. An outcome-led aid worker might hesitate to feed someone they suspect of being a mass murderer, in case he or she is able to commit further crimes.

Many non-governmental organizations (NGOs) which began as welfare or development organizations hold to this tradition, believing it is legitimate to address the cause of suffering and the consequences of actions even if this may limit their ability to alleviate suffering whenever and wherever it is found.

While it is possible to debate which approach is right or better, the problem is that many in aid try to practice both – development agencies take on relief work and humanitarian agencies are unable to ignore the causes of suffering. It is not

---

Box 11.1 **Minimal performance standards**

In late 1996, the Steering Committee for Humanitarian Response (SCHR) and the US NGO coalition InterAction agreed to begin a joint one-year programme to develop a set of "common minimal performance standards" in humanitarian response and a beneficiaries' charter.

The project recognizes "the continuing need to improve the quality of assistance provided by NGOs and the International Federation of Red Cross and Red Crescent Societies to those caught up in disasters, and to improve the accountability of agencies to their membership, their donors and their beneficiaries".

Standards will be based on the recognition and elaboration of a set of rights – derived from existing law and conventions – relevant to all beneficiaries in disaster situations. Drawing on existing material and current best-practice, the standards will cover minimal and relative entitlements, the delivery of assistance

and the accountability for action. Where necessary, new standards will be drafted where no suitable ones presently exist, and all the rights pertinent to disaster victims will be highlighted in a charter statement.

The charter and performance standards will be developed in collaboration with leading NGOs, interested donor governments and UN agencies to ensure final acceptability and a high degree of ownership. All major donors currently funding humanitarian assistance will be encouraged to contribute to the project and will be kept appraised of its development.

As with the *Code of Conduct*, SCHR and InterAction members will first commit to applying these standards to their own work, but will then actively encourage their formal adoption and practice across the humanitarian response community by relief agencies and their donors.

---

impossible to have both approaches in one organization, just as a government has ministries of both defence and health, but it is profoundly wrong to mix the two operationally in situations where lives are at risk.

It is wrong for two reasons. First, practising a partisan approach while claiming to be neutral in delivering relief risks jeopardizing the ability of the agency to actually deliver goods where and when they are needed. Thus the beneficiaries of the agency suffer. Secondly, in most disaster situations, the actions of one NGO very quickly reflect on the credibility of all others. If one agency tries to combine both approaches, it may jeopardize the ability of every organization to deliver assistance.

While outcome-led agencies can engage in the often difficult and messy business of deciding whose "side" they are on, humanitarian agencies cannot make this choice. It is about duty and rights not outcomes, absolute values not relative ones, the sanctity of the individual not the well-being of the community. Those absolutes can make hard choices easier to take, from whether the military should do humanitarian work to debates over developmental relief.

By definition, foreign military forces are not independent, and often not neutral. They represent the forceful expression of political will. Linking them to humanitarian work removes its shield of neutrality and independence, jeopardizing the ability to alleviate suffering now and in the future. There is nothing wrong with military forces working in the same geographical area as humanitarians. An impartial military force can restore security and the rule of law for everyone, not just aid workers.

Amid the debate about the nature of relief, Operation Lifeline Sudan (OLS) has tended towards a developmental approach in recent years, placing relief in the context of the rights of the Sudanese state and Sudan's culture and economy.

In the words of the United Nations (UN) Development Programme's Resident Representative, quoted in the 1996 OLS evaluation: "We often define humanitarianism as putting bread in the mouth of a starving person, but it is not humanitarian to let him get into that situation. We should replace free food deliveries and make people repay what they have received, this is what we are doing....People should repay this humanitarian loan not to us but to their community.

"We are taking them out of the beggar mentality. People are proud to pay for themselves...this is part of society building, enabling people to become more consciously self-reliant. It is linked to democracy building because people have to elect a management committee."

Universal codes

While this may serve some future greater good, in the short term it erodes humanitarian standards. The evaluation has shown that the Sudanese get a lower quality and quantity of relief than the levels to which many believe they are entitled. Firstly, aid has been designed to adhere to the principle of sovereignty in a way some find of concern; second, it is tailored to achieve developmental aims which critics claim are hard to achieve in present circumstances.

Humanitarianism involves standards, which have to be defined, codified and described in an unambiguous way so people can claim them and so others can ensure that they are fulfilled. This is what International Humanitarian Law and human rights declarations seek to do at a general level. Missing is any expression of how these general rights translate into the day-to-day practice of those seeking to ensure their fulfilment. Hence the need for universal operational agency codes and minimal standards.

While 20 years ago states might have been expected to codify such standards, today operational agencies must act.

The power of the state is no longer all pervasive. For 100 years the "greater good", enshrined in the notion of the state, has gradually been eroded in favour of individual rights. The development of International Humanitarian Law already limits how states fight wars. The World Trade Organization and international trading conventions are limiting how states can protect their industry and markets. Human rights conditionality on aid attempts to circumscribe the power of the state.

The state is less and less the sole, or even the most important, guardian of the rights of the people. Other guardians include the indeterminate group of states known as the "international community", increasingly-influential community and citizen organizations, from consumer groups to the global environmental movement, and humanitarian agencies themselves.

While humanitarian agencies may derive their self-defined mandate from "the people", not the state, the relation between that mandate and the rights of the people is not so well defined.

A truer representation is ensured by states through the ballot box and due process of law. Agencies lack these mechanisms and must therefore develop others – codes and standards that guide humanitarian assistance – to keep themselves on the straight and narrow. The alternative is to have such mechanisms imposed on agencies by states through legislation or by beneficiaries and claimants through popular action.

The existing *Code of Conduct for the International Red Cross and Red Crescent Movement and NGOs in Disaster Relief* is a first attempt to do this. It is simple, laying

Box 11.2 Ten principal points in the *Code of Conduct*

The full text of the *Code of Conduct for the International Red Cross and Red Crescent Movement and NGOs in Disaster Relief* can be found in the *World Disasters Report 1994* in English, French, Spanish, Arabic, Finnish and Japanese, and on the International Federation's Internet Web site (http://www.ifrc.org).

The ten principal points that signatories have agreed to abide by are given below:

1. The humanitarian imperative comes first.

2. Aid is given regardless of race, creed or nationality of the recipients and without adverse distinction of any kind. Aid priorities are calculated on the basis of the need alone.

3. Aid will not be used to further a particular political or religious standpoint.

4. We shall endeavour not to act as instruments of government foreign policy.

5. We shall respect culture and custom.

6. We shall attempt to build disaster response on local capacities.

7. Ways shall be found to involve programme beneficiaries in the management of relief aid.

8. Relief aid must strive to reduce future vulnerabilities to disaster as well as meeting basic needs.

9. We hold ourselves accountable to both those we seek to assist and those from whom we accept resources.

10. In our information, publicity and advertising activities, we shall recognize disaster victims as dignified humans, not hopeless objects.

The *Code of Conduct* Register

The International Federation is keeping a public record of all NGOs who register their commitment to the *Code of Conduct*, and will publish the list periodically in the *World Disasters Report*. The full text of the Code, including a registration form, is published by the International Federation as a short booklet in English, French, Spanish and Arabic, and is available upon request. A 10-minute, *Code of Conduct* training video with notes is also available – in the same four languages as the booklet – at a small charge.

For details, write to: *Code of Conduct*, International Federation of Red Cross and Red Crescent Societies, PO Box 372, 1211 Geneva 19, Switzerland, tel.: (41)(22) 730 4222; fax: (41)(22) 733 0395; e-mail: walker@ifrc.org

down ten basic principles of behaviour for agencies and their staff and then outlines what, in turn for keeping to these principles, agencies expect from the state.

The Code has now been accepted by more than 100 independent humanitarian agencies. At the 1995 International Red Cross Conference, 144 state signatories to the Geneva Conventions welcomed the Code and agreed to encourage NGOs in their countries to use it. The UN High Commissioner for Refugees, UN Children's Fund and World Food Programme are considering how to incorporate the Code into their criteria for partner agencies. Donor agencies, such as the Danish DANIDA, Swedish SIDA and the British Overseas Development Administration, are considering whether Code compliance should be a funding criteria.

Since the Code is essentially about agencies' behaviour and not about direct service to beneficiaries, there is a clear need to go further and establish universal operational standards.

The past five years have seen a leap in the demand for humanitarian relief, a proliferation of agencies offering assistance, an increased degree of competition between agencies, a growing flow of funding from government sources into independent relief agencies, and a concern that demand for humanitarian assistance is outstripping available resources.

The humanitarian system must now, more than ever before, demonstrate effective resource allocation and efficient resource utilization. It must be able to make a more

---

Box 11.3 **People in Aid: best practice for aid workers**

The people who work for aid agencies are their most valuable resource, turning the values of humanitarian relief or development into action.

That critical importance has been recognized in the People in Aid Code of Best Practice in the Management and Support of Aid Personnel, launched in early 1997.

People in Aid is a group of 11 UK-based international aid organizations, including the British Red Cross, major NGOs and the government Overseas Development Administration. The code reflects their concerns about the vulnerability of aid workers and their commitment to effective aid, good management and staff protection.

The code has detailed information, examples of good practice and sets of key indicators under these seven principles:

1. The people who work for us are integral to our effectiveness and success.
2. Our human resources policies aim for best practice.
3. Our human resources policies aim to be effective, efficient, fair and transparent.
4. We consult our field staff when we develop human resources policy.
5. Plans and budgets reflect our responsibilities towards our field staff.
6. We provide appropriate training and support.
7. We take all reasonable steps to ensure staff security and well being.

A broad range of aid agencies are now participating in a pilot implementation process, assisted by People in Aid and leading to accreditation by an independent human resources auditor.

---

objective and coherent case for additional resources when these are required. It is time to move forward with a technical elaboration of the Code.

Agencies have to be able to lay down, unambiguously, what beneficiaries have a right to expect in terms of what is delivered or secured, and how it is provided. The urgency of this need was most clearly articulated in the 1996 multi-donor evaluation of the Rwanda crisis, which factually set out the beneficiary suffering caused by poor agency performance standards, demonstrating the need for universal standards and a monitoring mechanism for agency quality.

While much work has already been done in defining minimum standards and good practice, the impact and usefulness of this work has been limited by the lack of agreement over how these should operate or be applied. Agencies have been reluctant to commit themselves to standards which may, for quite logical reasons, be impossible to meet. However, this should not prevent agencies providing a transparent commitment to "state of the art" humanitarian standards. That these will change over time is no justification for not establishing standards in the first instance. Such a set of guidelines should cover the four essential sectors of relief assistance: food and nutrition; water and sanitation; medical care; clothing, shelter and settlements, including the selection of relief camp sites.

Many previous attempts at deriving standards have focused exclusively on the end point of assistance delivery: what could be called the entitlements, such as food rations. This work to establish entitlement norms, to which agencies are publicly committed, must be completed. After this first step, agencies must set down guidance for how those entitlements are delivered, covering issues such as local procurement, targeting and distribution systems.

Agencies have to be equally concerned with actions after delivery. Agencies should be accountable to their beneficiaries, to themselves through programme monitoring, to donors and, through evaluation, to future programmes. Guidelines need to cover:

■ What the agency should deliver, or ensure is available, as a minimum for survival, i.e., what is needed to fulfil the absolute minimum entitlement of the disaster victim.

■ How relief will be made available to beneficiaries.

■ Forms of accountability to which the agency aspires, to beneficiaries and the local population, to donors, to its own staff and membership, and to future operations (in the form of evaluations and a commitment to continuous improvement).

■ Cross-cutting issues, such as environment and gender.

A set of standards has to be built on what people have a right to, not what agencies, donors or host states feel is possible. For that reason, there is a need to extract and highlight clauses relevant to the rights of disaster victims from existing international law and declarations.

Production of a beneficiaries' charter, offering clear statements – derived from existing accepted universal rights – would provide the basis for a great improvement in transparency and accountability of the humanitarian system. It could also facilitate the emergence of organizations of people affected by humanitarian emergencies.

Humanitarian work is more complex today. Agencies face increasing pressures to compromise the principles upon which most were founded. And, if these principles are compromised, those affected by disaster often end up with a poorer service. Clarity over the ethics of each agency and adherence to universally-agreed standards of practice are essential if humanitarians are to ensure continued improvement in the quality of the service offered to disaster victims.

Sources, references and further information

Minear, Larry and Weiss, Thomas G. *Humanitarian Action in Times of War; A Handbook for Practitioners.* Boulder and London: Lynne Rienner Publishers, 1995.

ODI. *The People in Aid Code of Best Practice in the Management and Support of Aid Personnel.* Network Paper 20, Relief and Rehabilitation Network, Overseas Development Institute, London, 1997.

Web sites

One World: http://www.oneworld.org

Relief and Rehabilitation Network: http://www.oneworld.org/odi/rrn/index.html

Humanitarianism and War Project: http://www.brown.edu:80/departments/Watson_Institute/H_W/

As of 1 March 1997, 113 agencies have registered their commitment to the Code.

Australia	Care Australia
Austria	Austrian Relief Programme (ARP)
Bangladesh	Youth Approach for Development and Cooperation
Belgium	Agora – Vitrine du Monde
Belgium	Care International
Belgium	Centre International de Formation des Cadres du Développement (C.I.F.C.D.)
Belgium	Institute of Cultural Affairs International – ZAGREB
Belgium	Oxfam
Benin	Conseil des Activités Educatives du Bénin
Canada	Adventist Development and Relief Agency (ADRA)
Canada	Family to Family
Canada	Oxfam
Côte d'Ivoire	ADRA
Croatia	ADEH International
Croatia	Pax Christi (Germany)
Denmark	ADRA
Denmark	Dan Church Aid
Dominica	Brisin Agencies, Ltd.
Dominica	Dominica Christian Council
Dominica	Society of St Vincent de Paul
Ethiopia	Selam Children's Village
Finland	Save the Children
France	ADRA
France	Enfants du Monde
France	Enfants Réfugiés du Monde
Germany	ADRA
Greece	Institute of International Social Affairs
Guinea	Commission Africaine des Promoteurs de la Santé et des Droits de l'Homme (CAPSDH)
Hong Kong	Oxfam

India	ADRA
India	Ambiha Charitable Trust
India	ASHA (Action for Social & Human Acme)
India	Centre for Research on Ecology, Environmental Education, Training and Education (CREATE)
India	Federation of Interfaith Orphanage and Allied Educational Relief Technical Training Institutions
India	Institute for Youth and Disaster Preparedness
India	Mahila Udyamita Vikas Kalya Evan Siksha Sansthah
Ireland	Express Aid International
Italy	Associazione Amici dei Bambini
Italy	Caritas Internationalis
Italy	Centro Internazionale di Cooperazione allo Sviluppo
Italy	Comitato Collaborazione Medica (CCM)
Italy	Comitato di Coordinamento delle Organizzazioni per il Servizio Volontario
Italy	International College for Health Cooperation in Developing Countries (CUAMM)
Italy	Movimondo
Italy	Reggio Terzo Mondo (R.T.M.)
Italy	Volontari Italiani Solidarietà Paesi Emergenti
Japan	Association of Medical Doctors of Asia (AMDA)
Laos	ADRA
Lebanon	Disaster Control Centre
Luxembourg	Amicale Rwanda-Luxembourg
Myanmar	ADRA
Netherlands	Caritas Nederland
Netherlands	Disaster Relief Agency
Netherlands	Dorcas Aid International
Netherlands	Dutch Interchurch Aid
Netherlands	Netherlands Organisation for International Cooperation (NOVIB)
Netherlands	Tear Fund

Netherlands	Terre des Hommes		UK	ADRA, Trans-Europe
Netherlands	ZOA Refugee Care		UK	Children in Crisis
New Zealand	Oxfam		UK	Children's Aid Direct
New Zealand	Tear Fund		UK	Hope and Homes for Children
Norway	Norwegian Organisation for Asylum Seekers		UK	Human Appeal International
Norway	Norwegian Refugee Council		UK	International Extension College
Philippines	ADRA		UK	Marie Stopes International
Portugal	Instituto Portugues de Medicina Preventiva		UK	Medical Emergency Relief International (MERLIN)
Russia	ADRA, Euro-Asia Division		UK	Oxfam
Spain	Intermón		UK	Post-War Reconstruction and Development Unit
Spain	Radioaficionados Sin Fronteras		UK	RedR
Sri Lanka	ADRA		UK	Save the Children Fund
Swaziland	Save the Children Fund		UK	Tear Fund
Sweden	Swedish Fellowship of Reconciliation (SWEFOR)/Kristna Fredsrorelsen		UK	The Ockenden Venture
			UK	The Salvation Army
Sweden	The Swedish Organisation for Individual Relief		UK	UK Foundation for the Peoples of the South Pacific
Switzerland	Ananda Marga Universal Relief Team (AMURT)		UK	World Association of Girl Guides and Girl Scouts
Switzerland	Association for the Children of Mozambique		USA	International Medical Corps
Switzerland	Catholic Relief Services		USA	International Rescue Committee
Switzerland	Commission Internationale Catholique pour les Migrations		USA	Lutheran World Relief
Switzerland	Food for the Hungry International		USA	Operation USA
Switzerland	Interaid International		USA	Oxfam
Switzerland	International Committee of the Red Cross		USA	Truck Aid International
Switzerland	International Federation of Red Cross and Red Crescent Societies		USA	Women's Commission for Refugee Women and Children
Switzerland	International Save the Children Alliance		Zaire	Humanitas, Corps de Sauvetage
Switzerland	Lutheran World Federation		Zaire	Oxfam
Switzerland	World Council of Churches			
Switzerland	World Vision International			
Thailand	ADRA			
UK	Action Against Hunger			
UK	Actionaid			

Chapter

12 National Societies: working as a Federation

Contact details for the members of the International Red Cross and Red Crescent Movement.

THE INTERNATIONAL RED CROSS AND RED CRESCENT MOVEMENT

The International Federation of Red Cross and Red Crescent Societies

P.O. Box 372
1211 Geneva 19
SWITZERLAND
Tel. (41)(22) 730 42 22
Fax (41)(22) 733 03 95
Tlx (045) 412 133 FRC CH
Tlg. LICROSS GENEVA
E-mail secretariat@ifrc.org
WWW http://www.ifrc.org

The International Committee of the Red Cross

19 avenue de la Paix
1202 Geneva
SWITZERLAND
Tel. (41)(22) 734 60 01
Fax (41)(22) 733 20 57
Tlx 414 226 CCR CH
Tlg. INTERCROIXROUGE GENEVE
E-mail icrc.gva@gwn.icrc.org
WWW http://www.icrc.org

NATIONAL RED CROSS AND RED CRESCENT SOCIETIES

National Red Cross and Red Crescent Societies are listed alphabetically by International Organization for Standardization Codes for the Representation of Names of Countries, English spelling.

Details correct as of 1 March 1997. Please forward any corrections to the International Federation's Information Resource Centre in Geneva (e-mail: irc@ifrc.org).

Afghan Red Crescent Society

Puli Hartan
Kabul
Postal Address:
P.O. Box 3066
Shahre - Naw Kabul
AFGHANISTAN
Tel. 32357 / 32211
Tlx arcs af 24318
Tlg. SERAMIASHT KABUL

Albanian Red Cross

Rruga "Muhammet Gjollesha"
Sheshi "Karl Topia"
Tirana
Postal Address: C.P. 1511
Tirana
ALBANIA
Tel. (355)(42) 25855; 22037
Fax (355)(42) 25855
Tlg. ALBCROSS TIRANA

Algerian Red Crescent

15 bis, Boulevard
Mohammed V
Alger
ALGERIA
Tel. (213) 2645727;
2645728;
Fax (213) 2649787
Tlx 56056 HILAL ALGER
Tlg. HILALAHMAR ALGER

Andorra Red Cross

Prat de la Creu 22
Andorra la Vella
ANDORRA
Tel. (376) 825225
Fax (376) 828630
Tlx AND 208 "Att. CREU ROJA"

Angola Red Cross

Rua 1 Congresso no 21
Luanda
Postal Address:
Caixa Postal 927
Luanda
ANGOLA
Tel. (244)(2) 336543;
333991
Fax (244)(2) 345065
Tlx 3394 CRUZVER AN

Antigua and Barbuda Red Cross Society

Red Cross House
Old Parham Road
St. Johns, Antigua W.I.
Postal Address:
P.O. Box 727
St. Johns, Antigua W.I.
ANTIGUA AND BARBUDA
Tel. (1)(809) 4620800
Fax (1)(809) 4620800
Tlx 2195 DISPREP "For Red Cross"

Argentine Red Cross

Hipólito Yrigoyen 2068
1089 Buenos Aires
ARGENTINA
Tel. (54)(1) 9511391;
9511854
Fax (54)(1) 9527715
Tlx 21061 CROJA AR
Tlg. ARGENCROSS
BUENOS AIRES

Armenian Red Cross Society

Antarain str. 188
Yerevan 375019
ARMENIA
Tel. (374)(2) 560630;
583630
Fax (374)(2) 583630
Tlx 243345 ODER SU,
Country code 64

Australian Red Cross

206 Clarendon Street
East Melbourne, Vic 3002
Postal Address:
P.O. Box 100
East Melbourne Vic 3002
AUSTRALIA
Tel. (61)(3) 94185200
Fax (61)(3) 94190404
E-mail
arcnatnl@c031.aone.net.au

Austrian Red Cross

Wiedner Hauptstrasse 32
Wien 4
Postal Address: Postfach 39
1041 Wien 4
AUSTRIA
Tel. (43)(1) 58900-0
Fax (43)(1) 58900-199
Tlx oerk a 133111
Tlg. AUSTROREDCROOS
WIEN
E-mail oerk@redcross.or.at

Red Crescent Society of Azerbaijan

Prospekt Azerbaidjan 19
Baku
AZERBAIJAN
Tel. (994)(12) 931912;
938481
Fax (994)(12) 931578

The Bahamas Red Cross Society

John F. Kennedy Drive
Nassau
Postal Address:
P.O. Box N-8331
Nassau
BAHAMAS
Tel. (1)(809) 3237370;
3237371
Fax (1)(809) 3237404
Tlx 20657 BAHREDCROSS
Tlg. BAHREDCROSS
NASSAU

Bahrain Red Crescent Society

P.O. Box 882
Manama
BAHRAIN
Tel. (973) 293171
Fax (973) 291797
Tlg. HILAHAMAR MANAMA

Bangladesh Red Crescent Society

684-686 Bara Maghbazar
Dhaka - 1217
Postal Address:
G.P.O. Box 579
Dhaka
BANGLADESH
Tel. (880)(2) 407908; 406902
Fax (880)(2) 831908
Tlx 632232 BDRC BJ
Tlg. RED CRESCENT
DHAKA

The Barbados Red Cross Society

Red Cross House
Jemmotts Lane
Bridgetown
BARBADOS
Tel. (1)(809) 4262052
Fax (1)(809) 426-2052 "For Red Cross"
Tlx 2201 P.U.B. T.L.X. W.B.
Tlg. REDCROSS
BARBADOS

Belarusian Red Cross

35, Karl Marx Str.
220030 Minsk
BELARUS
Tel. (375)(17) 2272620
Fax (375)(17) 2272620
Tlx 252290 KREST SU

Belgian Red Cross

Ch. de Vleurgat 98
1050 Bruxelles
BELGIUM
Tel. (32)(2) 6454411
(French); 6454458 (Flemish)
Fax (32)(2) 6460439
(French); 6460441 (Flemish)
Tlx 24266 BELCRO B
Tlg. CROIXROUGE
BELGIQUE BRUXELLES
E-mail
belgian.redcross@infoboard.
be

Belize Red Cross Society

1 Gabourel Lane
Belize City
Postal Address:
P.O. Box 413
Belize City
BELIZE
Tel. (501)(2) 73319
Fax (501)(2) 30998
Tlx BTL BOOTH 211 Bze
attn. Red Cross

Red Cross of Benin

B.P. No. 1
Porto-Novo
BENIN
Tel. (229) 212886
Fax (229) 214927
Tlx 1131 CRBEN

Bolivian Red Cross

Avenida Simón Bolívar
N 1515
La Paz
Postal Address:
Casilla No. 741
La Paz
BOLIVIA
Tel. (591)(2) 340948;
326568
 Fax (591)(2) 359102
Tlx 2220 BOLCRUZ
Tlg. CRUZROJA - LA PAZ

Botswana Red Cross Society

135 Independance Avenue
Gaborone
Postal Address:
P.O. Box 485
Gaborone
BOTSWANA
Tel. (267) 352465; 312353
Fax (267) 312352
Tlg. THUSA GABORONE

Brazilian Red Cross

Praça Cruz Vermelha No. 10
20230-130 Rio de Janeiro
RJ
BRAZIL
Tel. (55)(21) 2323223;
2326325
Fax (55)(21) 2426760
Tlx (38) 2130532 CVBR BR
Tlg. BRAZCROSS RIO DE
JANEIRO

Brunei Darussalam Red Crescent Society*

P.O. Box 3065
Bandar Seri Begawan 1930
BRUNEI DARUSSALAM
Tel. (673)(2) 339774
Fax (673)(2) 339572

*Will be definitively admitted
by the General Assembly in
November 1997*

Bulgarian Red Cross

61, Dondukov Boulevard
1527 Sofia
BULGARIA
Tel. (359)(2) 441443;
441444
Fax (359)(2) 441759
Tlx 23248 B CH K BG
Tlg. BULGAREDCROSS
SOFIA

Burkinabe Red Cross Society

01 B.P. 4404
Ouagadougou 01
BURKINA FASO
Tel. (226) 300877
Fax (226) 363121
Tlx LSCR 5438 BF
OUAGADOUGOU

Burundi Red Cross

18, Av. de la Croix-Rouge
Bujumbura
Postal Address: B.P. 324
Bujumbura
BURUNDI
Tel. (257) 216246; 211101
Tlx 5081 CAB PUB BDI

Cambodian Red Cross Society

17, Vithei de la Croix-Rouge
Cambodgienne
Phnom-Penh
CAMBODIA

Cameroon Red Cross Society

Rue Henri Dunant
Yaoundé
Postal Address: B.P. 631
Yaoundé
CAMEROON
Tel. (237) 224177
Fax (237) 224177
Tlx (0970) 8884 KN

The Canadian Red Cross Society

1800 Alta Vista Drive
Ottawa, Ontario KIG 4J5
CANADA
Tel. (1)(613) 7393000
Fax (1)(613) 7311411
Tlx CANCROSS 05-33784
Tlg. CANCROSS OTTA-
WA
E-mail can-
cross@redcross.ca

Red Cross of Cape Verde

Rua Andrade Corvo
Caixa Postal 119
Praia
CAPE VERDE
Tel. (238) 611701; 614169
Fax (238) 614174
Tlx 6004 CV CV

Central African Red Cross Society

Avenue Koudoukou Km, 5
Bangui
Postal Address: B.P. 1428
Bangui
CENTRAL AFRICAN
REPUBLIC
Tel. (236) 612223
Fax (236) 613561
Tlx DIPLOMA 5213 "Pour
Croix-Rouge"

Red Cross of Chad

B.P. 449
N'djamena
CHAD
Tel. (235) 515218; 5 13434
Tlg. CROIXROUGE
N'DJAMENA

Chilean Red Cross

Avenida Santa María No.
150 Providencia
Santiago de Chile
Postal Address: Correo 21,
Casilla 246 V
Santiago de Chile
CHILE
Tel. (56)(2) 7771448
Fax (56)(2) 7370270
Tlx 340260 PBVTR CK
Tlg. "CHILECRUZ"

Red Cross Society of China

53 Ganmian Hutong
100010 Beijing
CHINA
Tel. (86)(10) 65124447;
65135838
Fax (86)(10) 65124169
Tlx 210244 CHNRC CN
Tlg. HONGHUI BEIJING
E-mail
xiaox9@hns.cjfh.ac.cn

Colombian Red Cross Society

Avenida 68 N 66-31
Santafé de Bogotá D.C.
Postal Address: Apartado
Aéreo 11-10
Bogotá, D.C.
COLOMBIA
Tel. (57)(1) 2506611
Fax (57)(1) 2319208
E-mail
scrcol@col1.telecom.com.co

Congolese Red Cross

Place de la Paix
Brazzaville
Postal Address: B.P. 4145
Brazzaville
CONGO
Tel. (242) 824410
Fax (242) 828825
Tlx UNISANTE 5364 Pour
"Croix-Rouge"

Costa Rican Red Cross

Calle 14, Avenida 8
San José 1000
Postal Address:
Apartado 1025
San José 1000
COSTA RICA
Tel. (506) 2337033; 2553761
Fax (506) 2237628
Tlx 2547 COSTACRUZ SAN
JOSÉ
Tlg. COSTACRUZ SAN
JOSÉ

Red Cross Society of Côte d'Ivoire

P.O. Box 1244
Abidjan 01
COTE D'IVOIRE
Tel. (225) 321335
Fax (225) 225355
Tlx 24122 SICOGI CI

Croatian Red Cross

Ulica Crvenog kriza 14
10000 Zagreb
CROATIA
Tel. (385)(1) 415458;
415469
Fax (385)(1) 4550072
E-mail redcross@hck.hr

Cuban Red Cross

Calle Calzada No. 51
Vedado
 Ciudad Habana
C.P. 10400
CUBA
Tel. (53)(7) 326005
Fax (53)(7) 662057
Tlx 511149 MSP CU para
Cruz Roja
Tlg. CRUROCU HABANA

Czech Red Cross

Thunovska 18
CZ-118 04 Praha 1
CZECH REPUBLIC
Tel. (42)(2) 245 103 47;
245 103 18
Fax (42)(2) 245 103 18
Tlx 122 400 csrc c
Tlg. CROIX PRAHA

Danish Red Cross

Blegdamsvej 27
2100 København Ö
Postal Address:
P.O. Box 2600
DK-2100 København Ö
DENMARK
Tel. (45) 35259200
Fax (45) 31421186
Tlx 15726 DANCRO DK
Tlg. DANCROIX
KÖBENHAVN
E-mail drc@redcross.dk

Red Crescent Society of Djibouti

B.P. 8
Djibouti
DJIBOUTI
Tel. (253) 352451
Fax (253) 355049
Tlx 5871 PRESIDENCE DJ

Dominica Red Cross Society

Federation Drive
Goodwill
DOMINICA
Tel. (1)(809) 4488280
Fax (1)(809) 4487708
Tlx   8625 TELAGY DO -
for Dominica RC
Tlg. DOMCROSS

Dominican Red Cross

Calle Juan E. Dunant No. 51
Ens. Miraflores
Santo Domingo
Postal Address:
Apartado Postal 1293
Santo Domingo, D.N.
DOMINICAN REPUBLIC
Tel. (1)(809) 6823793;
6897344
Fax (1)(809) 6822837
Tlx rca sdg 4112 "PARA
CRUZ ROJA DOM."
 Tlg. CRUZ ROJA
DOMINICANA, SANTO
DOMINGO
E-mail
cruz.roja@codetel.net.do

Ecuadorian Red Cross

Av. Colombia y Elizalde Esq.
Quito
Postal Address:
Casilla 17-01-2119
Quito
ECUADOR
Tel. (593)(2) 514587
Fax (593)(2) 570424
Tlx CRUZRO 2662
Tlg. CRUZ ROJA QUITO
E-mail
crequito@uio.satnet.net

Egyptian Red Crescent Society

29, El Galaa Street
Cairo
EGYPT
Tel. (20)(2) 5750558;
5750397
Fax (20)(2) 5740450
Tlx 93249 ERCS UN
Tlg. 124 HELALHAMER

Salvadorean Red Cross Society

17 C. Pte. y Av. Henri
Dunant
San Salvador
Postal Address:
Apartado Postal 2672
San Salvador
EL SALVADOR
Tel. (503) 2227743;
2227749
Fax (503) 2227758
Tlx 20550 cruzalva
Tlg. CRUZALVA SAN SAL-
VADOR

Red Cross of Equatorial Guinea

Alcalde Albilio Balboa 92
Malabo
Postal Address:
Apartado postal 460
Malabo
EQUATORIAL GUINEA
Tel. (240)(9) 3701
Fax (240)(9) 3701
Tlx 099/1111 EG.PUB MBO
"Favor Transmetien Cruz
Roja Tel. 2393"

Estonia Red Cross

Lai Street 17
EE-0001 Tallinn
ESTONIA
Tel. (372) 6411643
Fax (372) 6411641
Tlx 173491

Ethiopian Red Cross Society

Ras Desta Damtew Avenue
Addis Ababa
 Postal Address:
P.O. Box 195
Addis Ababa
ETHIOPIA
Tel. (251)(1) 449364;
159074
Fax (251)(1) 512643
Tlx 21338 ERCS ET
Tlg. ETHIOCROSS
ADDISABABA
E-mail
ercs@padis.gn.apc.org

Fiji Red Cross Society

22 Gorrie Street
Suva
Postal Address:
GPO Box 569
Suva
FIJI
Tel. (679) 314133; 314138
Fax (679) 303818
Tlx 2279 Red Cross (Public
facility)
Tlg. REDCROSS SUVA

Finnish Red Cross

Tehtaankatu 1 a
FIN-00140 Helsinki
Postal Address:
P.O. Box 168
FIN-00141 Helsinki
FINLAND
Tel. (358)(9) 12931
Fax (358)(9) 1293352
Tlx 121331 FINCR FI
Tlg. FINCROSS HELSINKI
E-mail
forename.surname@redcross.fi

French Red Cross

1, Place Henry-Dunant
F-75384 Paris Cedex 08
FRANCE
Tel. (33)(1) 44431100
Fax (33)(1) 44431101
Tlx CR PARIS 642760 F
CRPAR
Tlg. CROIROUGE PARIS
086

The Gambia Red Cross Society

Kanifing - Banjul
Postal Address:
P.O. Box 472, Banjul
GAMBIA
Tel. (220) 392405; 393179
Fax (220) 394921
Tlx 2338 REDCROSS GV
Tlg. GAMREDCROSS
BANJUL

German Red Cross

Friedrich-Ebert-Allee 71
D-53113 Bonn
Postal Address:
Postfach 1460
 53004 Bonn
GERMANY
Tel. (49)(228) 5411
Fax (49)(228) 541290
Tlx 886619 DKRB D
Tlg. DEUTSCHROTKREUZ
BONN
E-mail drk@drk.de

Ghana Red Cross Society

Ministries Annex Block A3
Off Liberia Road Extension
Accra
Postal Address:
P.O. Box 835
Accra
GHANA
Tel. (233)(21) 662298
Fax (233)(21) 232133
Tlx 2655 GRCS GH
Tlg. GHANACROSS ACCRA

Hellenic Red Cross

Rue Lycavittou 1
Athens 106 72
GREECE
Tel. (30)(1) 3646005;
3628648
Fax (30)(1) 3613564
Tlx 225156 EES GR
Tlg. HELLECROIX
ATHENES

Grenada Red Cross Society

Upper Lucas Street
St. George's
Postal Address:
P.O. Box 551
St. George's
GRENADA
Tel. (1)(809) 4401483
Fax (1)(809) 4401483

Guatemalan Red Cross

3a Calle 8 - 40, Zona 1
Guatemala, C.A.
GUATEMALA
Tel. (502)(2) 532026;
532027
Fax (502)(2) 324649
Tlx 5366 CROJA GU
Tlg. GUATECRUZ
GUATEMALA

Red Cross Society of Guinea

B.P. 376
Conakry
GUINEA
Tel. (224) 443825
Fax (224) 414255
Tlx 22101

Red Cross Society of Guinea-Bissau

Avenida Unidade Africana,
No. 12
Bissau
 Postal Address: Caixa
postal 514-1036 BIX, Codex
Bissau
GUINEA-BISSAU
Tel. (245) 212405
Tlx   251 PCE BI

The Guyana Red Cross Society

Eve Leary
Georgetown
Postal Address:
P.O. Box 10524
Georgetown
GUYANA
Tel. (592)(2) 65174
Fax (592)(2) 66523
Tlx 2226 FERNA GY "For
Guyana Red Cross"
Tlg. GUYCROSS
GEORGETOWN

Haitian National Red Cross Society

1, rue Eden
Bicentenaire
Port-Au-Prince
Postal Address: CRH
B.P. 1337
Port-Au-Prince
HAITI
Tel. (509) 225553; 225554
Fax (509) 231054
Tlg. HAITICROSS PORT
AU PRINCE

Honduran Red Cross

7a Calle
entre 1a. y 2a. Avenidas
Comayagüela D.C.
HONDURAS
Tel. (504) 378876; 374628
Fax (504) 380185
Tlx 1437 CRUZ R HO
Tlg. HONDUCRUZ
COMAYAGUELA

Hungarian Red Cross

Arany János utca 31
1051 Budapest V.
Postal Address: Magyar
Vöröskereszt
1367 Budapest 5, Pf. 121
HUNGARY
Tel. (36)(1) 1313950;
1317711
Fax (36)(1) 1533988
Tlx 224943 REDCR H
Tlg. REDCROSS
BUDAPEST
E-mail mvk@slip.hrc.hu

Icelandic Red Cross

Raudararstigur 18
105 Reykjavik
Postal Address:
Postbox 5450
125 Reykjavik
ICELAND
Tel. (354) 5626722
Fax (354) 5623150
E-mail rki@centrum.is

Indian Red Cross Society

Red Cross Building
1 Red Cross Road
New Delhi 110001
INDIA
Tel. (91)(11) 3716441;
3716442
Fax (91)(11) 3717454
Tlx 3166115 IRCS IN
Tlg. INDCROSS NEW
DELHI

Indonesian Red Cross Society

Jl. Jenderal Datot Subroto
Kav. 96
Jakarta 12790
Postal Address:
P.O. Box 2009
Jakarta
INDONESIA
Tel. (62)(21) 7992325
Fax (62)(21) 7995188
Tlx 66170 MB PMI IA
Tlg. INDONCROSS JKT

Red Crescent Society of the Islamic Republic of Iran

Ostad Nejatolahi Ave.
Tehran
IRAN, ISLAMIC REPUBLIC
OF
Tel. (98)(21) 8849077;
8849078
Fax (98)(21) 8849079
Tlx 224259 RCIA-IR
Tlg. CROISSANT-ROUGE
TEHERAN

Iraqi Red Crescent Society

Al-Mansour
Baghdad
Postal Address:
P.O. Box 6143
Baghdad
IRAQ
Tel. (964)(1) 8862191;
5343922
Fax (964)(1) 8840872
Tlx 213331 HELAL IK
Tlg. REDCRESCENT
BAGHDAD

Irish Red Cross Society

16, Merrion Square
Dublin 2
IRELAND
Tel. (353)(1) 6765135;
6765136
Fax (353)(1) 6767171
Tlx 32746 IRCS EI
Tlg. CROSDEARG DUBLIN
E-mail redcross@iol.ie

Italian Red Cross

12, Via Toscana
 I - 00187 Roma
ITALY
Tel. (39)(6) 47591
Fax (39)(6) 4759223
Tlx 613421 CRIROM I
Tlg. CRIROM 00187

Jamaica Red Cross

Central Village
Spanish Town
St. Catherine
Postal Address:
76 Arnold Road
Kingston 5
JAMAICA West Indies
Tel. (1)(809) 98478602
Fax (1)(809) 9848272
Tlx COLYB JA 2397 "For
Red Cross"
Tlg. JAMCROSS
KINGSTON

Japanese Red Cross Society

1-3 Shiba Daimon,
1-Chome, Minato-ku
Tokyo-105
JAPAN
Tcl. (81)(3) 34381311
Fax (81)(3) 34358509
Tlx JARCROSS J 22420
Tlg. JAPANCROSS TOKYO
E-mail
rcjpn@ppp.bekkoame.or.jp

Jordan National Red Crescent Society

Madaba Street
Amman
Postal Address:
P.O. Box 10001
Amman 11151
JORDAN
Tel. (962)(6) 773141;
773142
Fax (962)(6) 750815
Tlx 22500 HILAL JO
Tlg. HALURDON AMMAN

Kenya Red Cross Society

Nairobi South "C"
(Belle Vue), off Mombasa
Road
Nairobi
Postal Address:
P.O. Box 40712
Nairobi
KENYA
Tel. (254)(2) 503781;
503789
Fax (254)(2) 503845
Tlx 25436 IFRC KE
Tlg. KENREDCROSS NAI-
ROBI

Red Cross Society of the Democratic People's Republic of Korea

Ryonwa 1, Central District
Pyongyang
 KOREA, DEMOCRATIC
PEOPLE'S REPUBLIC OF
Tel. (850)(2) 18111; 18222
Fax (850)(2) 3814644
Tlx 5355 DAEMUN KP
Tlg. KOREACROSS
PYONGYANG

The Republic of Korea National Red Cross

32 - 3ka, Namsan-dong
Choong-Ku
Seoul 100 - 043
KOREA, REPUBLIC OF
Tel. (82)(2) 7559301
Fax (82)(2) 7740735
Tlx ROKNRC K28585
Tlg. KORCROSS SEOUL

Kuwait Red Crescent Society

Al-Jahra St.
Shuweek
Postal Address:
P.O. Box 1359
13014 Safat
KUWAIT
Tel. (965) 4839114; 4815478
Fax (965) 4839114
Tlx 22729

Lao Red Cross

Avenue Sethathirath
Vientiane
Postal Address: B.P. 650
Vientiane
LAO PEOPLE'S
DEMOCRATIC REPUBLIC
Tel. (856)(21) 216610;
212036
Fax (856)(21) 215935
Tlx 4491 TE via PTT LAOS
Tlg. CROIXLAO VIENTIANE

Latvian Red Cross

1, Skolas Street
RIGA
LV-1010
LATVIA
Tel. (371)(7) 310902
Fax (371)(7) 310902

Lebanese Red Cross

Rue Spears
Beyrouth
LEBANON
Tel. (961)(1) 372801;
372802
Fax (961)(1) 863299
Tlx CROLIB 20593 LE
Tlg. LIBACROSS
BEYROUTH

Lesotho Red Cross Society

23 Mabile Road
Maseru 100
Postal Address:
.O. Box 366
Maseru 100
LESOTHO
Tel. (266) 313911
Fax (266) 310166
Tlx 4515 LECROS LO
Tlg. LESCROSS MASERU
E-mail lrcs@wn.apc.org

Liberian Red Cross Society

107 Lynch Street
1000 Monrovia 20
Postal Address:
P.O. Box 20-5081
1000 Monrovia 20
LIBERIA
Tel. (231) 225172
Tlx 44210

Libyan Red Crescent

P.O. Box 541
Benghazi
LIBYAN ARAB JAMAHIRIYA
Tel. (218)(61) 9095827;
9099420
Fax (218)(61) 9095829
Tlx 40341 HILAL PY
Tlg. LIBHILAL BENGHAZI

Liechtenstein Red Cross

Heiligkreuz 25
FL-9490 Vaduz
LIECHTENSTEIN
Tel. (41)(75) 2322294
Fax (41)(75) 2322240
Tlg. ROTESKREUZ VADUZ

Lithuanian Red Cross Society

Gedimino ave. 3a
2600 Vilnius
LITHUANIA
Tel. (370)(2) 628947;
611914
Fax (370)(2) 619923

Luxembourg Red Cross

Parc de la Ville
L - 2014 Luxembourg
Postal Address: B.P. 404
L - 2014 LUXEMBOURG
Tel. (352) 450202; 450201
Fax (352) 457269
Tlg. CROIXROUGE
LUXEMBOURG

The Red Cross of The Former Yugoslav Republic of Macedonia

No. 13
Bul. Koco Racin
91000 Skopje
MACEDONIA, THE
FORMER YUGOSLAV
REPUBLIC OF
Tel. (389)(91) 114355
Fax (389)(91) 230 542

Malagasy Red Cross Society

1, rue Patrice Lumumba
Tsavalalana
Antananarivo
Postal Address: B.P. 1168
Antananarivo
MADAGASCAR
Tel. (261)(2) 22111
Fax (261)(2) 35457
Tlx 22248 COLHOT MG
"Pour Croix-Rouge"

Malawi Red Cross Society

Red Cross House (along
Presidential Way)
Lilongwe
Postal Address:
P.O. Box 30096
Capital City
Lilongwe 3
MALAWI
Tel. (265) 732877; 732878
Fax (265) 730210
Tlx 44276
E-mail
mrcs@unima.wn.apc.org

Malaysian Red Crescent Society

JKR 32, Jalan Nipah
Off Jalan Ampang
55000 Kuala Lumpur
MALAYSIA
Tel. (60)(3) 4578122;
4578236
Fax (60)(3) 4579867
Tlx MACRES MA 30166
Tlg. MALREDCRES KUALA
LUMPUR
E-mail mrcs@po.jaring.my

Mali Red Cross

Route Koulikoro
Bamako
Postal Address: B.P. 280
Bamako
MALI
Tel. (223) 224569
Fax (223) 220414
Tlx 2611 MJ

Malta Red Cross Society

104 St Ursola Street
Valletta 15400
MALTA
Tel. (356) 222645
Fax (356) 243664
E-mail
mltredcr@keyworld.net

Mauritanian Red Crescent

Avenue Gamal Abdel Nasser
Nouakchott
Postal Address: B.P. 344
Nouakchott
MAURITANIA
Tel. (222)(2) 51249
Fax (222)(2) 55686
 Tlx 5830 CRM

Mauritius Red Cross Society

Ste. Thérèse Street
Curepipe
MAURITIUS
Tel. (230) 6763604
Tlx YBRAT IW* 4258 "For
Mauritius Red Cross"
Tlg. MAUREDCROSS
CUREPIPE

Mexican Red Cross

Calle Luis Vives 200
Colonia Polanco
México, D.F. 11510
MEXICO
Tel. (52)(5) 3951111;
5800070
Fax (52)(5) 3951598
Tlx 01777617 CRMEME
Tlg. CRUZROJA MEXICO

Red Cross of Monaco

27, Boulevard de Suisse
Monte Carlo
MONACO
Tel. (377)(93) 506701
Fax (377)(93) 159047
Tlg. CROIXROUGE
MONTECARLO

Mongolian Red Cross Society

Central Post Office
Post Box 537
Ulaanbaatar
MONGOLIA
Tel. (976)(1) 320635
Fax (976)(1) 320934
Tlx 79358 MUIW
Tlg. MONRECRO

Moroccan Red Crescent

Palais Mokri
Takaddoum
Rabat
Postal Address: B.P. 189
Rabat
MOROCCO
Tel. (212)(7) 650898;
651495
Fax (212)(7) 759395
Tlx ALHILAL 319-40 M
RABAT
Tlg. ALHILAL RABAT

Mozambique Red Cross Society

Avenida 24 de Julho, 641
Maputo
Postal Address:
Caixa Postal 2986
Maputo
MOZAMBIQUE
Tel. (258)(1) 430045;
430046
Fax (258)(1) 429545
Tlx 6-169 CV MO

Myanmar Red Cross Society

Red Cross Building
42 Strand Road
Yangon
MYANMAR
Tel. (95)(1) 296552; 295238
Fax (95)(1) 296551
Tlx 21218 BRCROS BM
Tlg. MYANMARCROSS
YANGON

Namibia Red Cross

Red Cross House
100 Robert Mugabe Avenue
Windhoek
Postal Address:
P.O. Box 346
Windhoek
NAMIBIA
Tel. (264)(61) 235216;
235226
Fax (264)(61) 228949

Nepal Red Cross Society

Red Cross Marg
Kalimati
Kathmandu
Postal Address:
P.O. Box 217
Kathmandu
NEPAL
Tel. (977)(1) 270650;
270167
Fax (977)(1) 271915
Tlx 2569 NRCS NP
Tlg. REDCROSS
KATHMANDU
E-mail
nrcs@kalimati.mos.com.np

The Netherlands Red Cross

Leeghwaterplein 27
2521 CV The Hague
Postal Address:
P.O. Box 28120
2502 KC The Hague
NETHERLANDS
Tel. (31)(70) 4455666;
4455755
Fax (31)(70) 4455777
Tlx 32375 NRCS NL
Tlg. ROODKRUIS THE
HAGUE
E-mail hq@redcross.nl

New Zealand Red Cross

Red Cross House
14 Hill Street
Wellington 1
Postal Address:
P.O. Box 12140
Thorndon, Wellington
NEW ZEALAND
Tel. (64)(4) 4723750
Fax (64)(4) 4730315

Nicaraguan Red Cross

Reparto Belmonte
Carretera Sur
Managua
Postal Address:
Apartado 3279
Managua
NICARAGUA
Tel. (505)(2) 652082;
652084
Fax (505)(2) 651643
Tlx 2363 NICACRUZ
Tlg. NICACRUZ-MANAGUA
E-mail nicacruz@ibw.com.ni

Red Cross Society of Niger

B.P. 11386
Niamey
NIGER
Tel. (227) 733037; 722706
Fax (227) 723244
Tlx CRN GAP NI 5371

Nigerian Red Cross Society

11, Eko Akete Close
off St. Gregory's Road
South West Ikoyi
Lagos
Postal Address:
P.O. Box 764
Lagos
NIGERIA
Tel. (234)(1) 2695188;
2695189
Fax (234)(1) 680865
Tlx 21470 NCROSS NG
Tlg. NIGERCROSS LAGOS

Norwegian Red Cross

Hausmannsgate 7
0133 Oslo
Postal Address: Postbox 1.
Gronland
0133 Oslo
NORWAY
Tel. (47) 22054000
Fax (47) 22054040
Tlx 76011 NORCR N
Tlg. NORCROSS OSLO
E-mail
postman@redcross.no

Pakistan Red Crescent Society

Sector H-8
Islamabad
PAKISTAN
Tel. (92)(51) 854885;
856420
Fax (92)(51) 280530
Tlx 54103 PRCS PK
Tlg. HILALAHMAR
ISLAMABAD

Red Cross Society of Panama

Calle "E", N 11'50
Panamá
Postal Address:
Apartado 668
Zona 1 Panamá
PANAMA
Tel. (507) 2283014; 2280692
Fax (507) 2286857
Tlx 2661 STORTEXPA
Tlg. PANACRUZ PANAMA

Papua New Guinea Red Cross Society

Taurama Road
Port Moresby
Boroko
Postal Address:
P.O. Box 6545
Boroko
PAPUA NEW GUINEA
Tel. (675) 3258577; 3258759
Fax (675) 3259714
Tlx PNG RC NE 23292

Paraguayan Red Cross

Brasil 216 esq. José Berges
Asunción
PARAGUAY
Tel. (595)(21) 22797;
208199
Fax (595)(21) 211560
Tlg. CRUZ ROJA
PARAGUAYA
E-mail cruzroja@pla.net.py

Peruvian Red Cross

Av. Camino del Inca y
Nazarenas
Urb. Las Gardenias - Surco
Lima
Postal Address:
Apartado 1534
Lima
PERU
Tel. (51)(14) 482005;
489431
Fax (51)(14) 486472
Tlx 21002 -cp CESAR
Tlg. CRUZROJA PERUANA
LIMA

The Philippine National Red Cross

Bonifacio Drive
Port Area
Manila 2803
Postal Address:
PO Box 280
Manila 2803
PHILIPPINES
Tel. (63)(2) 5278384;
5278397
Fax (63)(2)5270857
Tlx 27846 PNRC PH
Tlg. PHILCROSS MANILA
E-mail
pnrcnhq@pdx.rpnet.com

Polish Red Cross

Mokotowska 14
00-561 Varsovie
Postal Address: P.O Box 47
00-950 Varsovie
POLAND
Tel. (48)(22) 6285201-7
Fax (48)(22) 6284168
Tlx 813561 PCK PL
Tlg. PECEKA VARSOVIE

Portuguese Red Cross

Jardim 9 de Abril, 1 a 5
1293 Lisboa Codex
PORTUGAL
Tel. (351)(1) 605571;
605650
Fax (351)(1) 3951045
Tlx 14369 PORCRS P
Tlg. CRUZVERMELHA

Qatar Red Crescent Society

P.O. Box 5449
Doha
QATAR
Tel. (974) 435111
Fax (974) 439950
Tlx 4753 qrcs dh
Tlg. hllal doha

Romanian Red Cross

Strada Biserica Amzei, 29
Sector 1
Bucarest
ROMANIA
Tel. (40)(1) 6593385;
6506233
Fax (40)(1) 3128452
Tlx 10531 romcr r
Tlg. ROMCROIXROUGE
BUCAREST

The Russian Red Cross Society

Tcheryomushkinski Proezd 5
117036 Moscow
RUSSIAN FEDERATION
Tel. (7)(095) 1266770;
1261403
Fax (7)(095) 2302867
Tlx 411400 IKPOL SU
Tlg. IKRESTPOL MOSKWA

Rwandan Red Cross

B.P. 425
Kigali
RWANDA
Tel. (250) 73302; 74402
Fax (250) 22558 UNHCR
"Pour C.R."
Tlx 22663 CRR RW

Saint Kitts and Nevis Red Cross Society

Red Cross House
Horsford Road
Basseterre
 Postal Address:
P.O. Box 62
Basseterre
SAINT KITTS AND NEVIS
Tel. (1)(809)(465) 2584
Fax (1)(809)(465) 2584

Saint Lucia Red Cross

Vigie
Castries St Lucia, W.I.
Postal Address:
P.O. Box 271
Castries St Lucia, W.I.
SAINT LUCIA
Tel. (1)(809) 4525582
Fax (1)(809) 45 37811
Tlx 6256 MCNAMARA LC
Attn. Mrs Boland

Saint Vincent and the Grenadines Red Cross

P.O. Box 431
SAINT VINCENT AND THE
GRENADINES
Tel. (1)(809) 4571816;
4571381
Fax (1)(809) 4572235
Tlx 7538 TA VQ "For Red
Cross"

Western Samoa Red Cross Society

P.O. Box 1616
Apia
SAMOA
Tel. (685) 23686
Fax (685) 22676
Tlx 779 224 MORISHED SX
(Attention Red Cross)

Red Cross of San Marino

Via Scialoja, Cailungo
Repubblica di San Marino,
47031
SAN MARINO
Tel. (37)(8) 994360
Fax (37)(8) 994360
Tlg. CROCE ROSSA
REPUBBLICA DI SAN
MARINO

Sao Tome and Principe Red Cross

Avenida 12 de Julho No.11
Sao Tomé
Postal Address: B.P. 96
Sao Tome
SAO TOME AND PRINCIPE
Tel. (239)(12) 22305; 22469
Fax (239)(12) 21365 publico
ST
Tlx 213 PUBLICO ST pour
"Croix-Rouge"

Saudi Arabian Red Crescent Society

General Headquarters
Riyadh 11129
SAUDI ARABIA
Tel. (966)(1) 4067956
Fax (966)(1) 4042541
Tlx 400096 HILAL SJ

Senegalese Red Cross Society

Boulevard F. Roosevelt
 Dakar
Postal Address: B.P. 299
Dakar
SENEGAL
Tel. (221) 233992
Fax (221) 225369
Tlx 61206 CSR SG

Seychelles Red Cross Society

B.P. 53
Mahe
SEYCHELLES
Tel. (248) 322122
Fax (248) 322122
Tlx 2302 HEALTH SZ

Sierra Leone Red Cross Society

6 Liverpool Street
Freetown
Postal Address:
P.O. Box 427
Freetown
SIERRA LEONE
Tel. (232)(22) 222384
Fax (232)(22) 229083
Tlx 3692 SLRCS
Tlg. SIERRA RED CROSS

Singapore Red Cross Society

Red Cross House
15 Penang Lane
Singapore 0923
SINGAPORE
Tel. (65) 3373587; 3360269
Fax (65) 3374360
Tlx SRCS RS 33978
Tlg. REDCROS
SINGAPORE

Slovak Red Cross

Grösslingova 24
814 46 Bratislava
SLOVAKIA
Tel. (42)(7) 325305; 323576
Fax (42)(7) 323279
Tlx 122 400 CSRC C
Tlg. CROIX PRAHA

Red Cross of Slovenia

Mirje 19
61000 Ljubljana
Postal Address:
P.O. Box 236
61111 Ljubljana
SLOVENIA
Tel. (386)(61) 1261200
Fax (386)(61) 1252142

The Solomon Islands Red Cross

P.O. Box 187
Honiara
 SOLOMON ISLANDS
Tel. (677) 22682
Fax (677) 25299
Tlx 66347 WING HQ

Somali Red Crescent Society

c/o ICRC Box 73226
Nairobi, Kenya
SOMALIA
Tel. Mogadishu (871 or 873)
131 2646; Nairobi (254)(2)
723963
Fax 1312647 (Mogadishu);
715598 (Nairobi)
Tlx 25645 ICRC KE

The South African Red Cross Society

25 Erlswold Way
Saxonwold
Johannesburg 2196
Postal Address:
P.O. Box 2829
Parklands 2121
SOUTH AFRICA
Tel. (27)(11) 4861313;
4861314
Fax (27)(11) 4861092
Tlg. REDCROSS
JOHANNESBURG

Spanish Red Cross

Rafael Villa, s/n (Vuelta
Ginés Navarro)
28023 El Plantío
Madrid
SPAIN
Tel. (34)(1) 3354444;
3354545
Fax (34)(1) 3354455
Tlx 23853 OCCRE E
Tlg. CRUZ ROJA
ESPANOLA MADRID

The Sri Lanka Red Cross Society

106, Dharmapala Mawatha
Colombo 7
Postal Address:
P.O. Box 375
Colombo
SRI LANKA
Tel. (94)(1) 691095; 699935
Fax (94)(1) 695434
Tlx 23312 SLRCS CE
Tlg. RED CROSS
COLOMBO

The Sudanese Red Crescent

P.O. Box 235
Khartoum
SUDAN
Tel. (249)(11) 772011
Fax (249)(11) 772877
Tlx 23006 LRCS SD
Tlg. EL NADJA KHARTOUM

Suriname Red Cross

Gravenberchstraat 2
Paramaribo
 Postal Address: Postbus
2919
Paramaribo
SURINAME
Tel. (597) 498410
Fax (597) 464780
Tlx 134 SURBANK "Att. Mr.
Linger for Red Cross"

Baphalali Swaziland Red Cross Society

104 Johnstone Street
Mbabane
Postal Address:
P.O. Box 377
Mbabane
SWAZILAND
Tel. (268) 42532
Fax (268) 46108
Tlx 2260 WD
Tlg. BAPHALALI MBABANE
E-mail bsrcs@wn.apc.org

Swedish Red Cross

Oesthammarsgatan 70
Stockholm
Postal Address: Box 27316
S-102 54 Stockholm
SWEDEN
Tel. (46)(8) 6655600
Fax (46)(8) 6612701
Tlx 19613 SWECROS S
Tlg. SWEDCROS
STOCKHOLM
E-mail
postmaster@redcross.se

Swiss Red Cross

Rainmattstrasse 10
3001 Bern
Postal Address: Postfach
3001 Bern
SWITZERLAND
Tel. (41)(31) 3877111
Fax (41)(31) 3877122
Tlx 911102 CRSB CH
Tlg. CROIXROUGE
SUISSE BERNE

Syrian Arab Red Crescent

Al Malek Aladel Street
Damascus
SYRIAN ARAB REPUBLIC
Tel. (963)(11) 4429662
Fax (963)(11) 4425677
Tlx 412857 HLAL
Tlg. CROISSANROUGE
DAMAS

Tanzania Red Cross National Society

Upanga Road
Dar es Salaam
Postal Address:
P.O. Box 1133
Dar es Salaam
TANZANIA, UNITED REPUBLIC OF
Tel. (255)(51) 24564; 24565
Tlx TACROS 41878
E-mail
redcrosstz@unidar.gn.apc.org

The Thai Red Cross Society

Paribatra Building
Central Bureau
1873, Rama IV Road
Bangkok-10330
THAILAND
Tel. (66)(2) 2564037;
2564038
Fax (66)(2) 2553727
Tlx 82535 threcso th
Tlg. THAICROSS
BANGKOK
E-mail
trcs@md2.md.chula.ac.th

Togolese Red Cross

51, rue Boko Soga
Amoutivé
Lome
Postal Address: B.P. 655
Lome
TOGO
Tel. (228) 212110
Fax (228) 215228
Tlx UNDERVPRO
5261/5145 "pour
Croix-Rouge"
Tlg. CROIX-ROUGE
TOGOLAISE LOME

Tonga Red Cross Society

P.O. Box 456
Nuku'Alofa
South West Pacific
TONGA
Tel. (676) 21360; 21670
Fax (676) 24158
Tlx 66222 CW ADM TS
Attn. Redcross
Tlg. REDCROSS TONGA

The Trinidad and Tobago Red Cross Society

Lot 7A, Fitz Blackman Drive
Wrightson Road
Port of Spain, West Indies
Postal Address: P.O. Box 357
Port of Spain
TRINIDAD AND TOBAGO
Tel. (1)(809) 6278215;
6278128
Fax (1)(809) 6278215
Tlx (294) 9003 for "Red Cross"
Tlg. TRINREDCROSS
PORT OF SPAIN

Tunisian Red Crescent

19, Rue d'Angleterre
Tunis 1000
TUNISIA
Tel. (216)(1) 240630;
245572
Fax (216)(1) 340151
Tlx 12524 HILAL TN
Tlg. HILALAHMAR TUNIS

Turkish Red Crescent Society

Genel Baskanligi
Karanfil Sokak No.7
06650 Kizilay
Ankara
TURKEY
Tel. (90)(312) 4317680
Fax (90)(312) 4177682
Tlx 44593 KZLY TR
Tlg. KIZILAY ANKARA

Red Crescent Society of Turkmenistan

48 A. Novoi str.
744000 Ashgabat
TURKMENISTAN
Tel. (7)(3632) 295512
Fax (7)(3632) 251750

The Uganda Red Cross Society

Plot 97, Buganda Road
Kampala
Postal Address: P.O. Box 494
Kampala
UGANDA
Tel. (256)(41) 258701;
258702
Fax (256)(41) 258184
Tlx (0988) 62118 redcrosug
Tlg. UGACROSS KAMPALA

Red Cross Society of Ukraine

30, ulitsa Pushkinskaya
252004 Kyiv
UKRAINE
Tel. (380)(44) 2250157;
2250334
Fax (380)(44) 2251096
Tlx 131329 LI RED/CROSS SU

Red Crescent Society of the United Arab Emirates

P.O. Box 3324
Abu Dhabi
UNITED ARAB EMIRATES
Tel. (971)(2) 661500
Fax (971)(2) 669919
Tlx 23582 RCS EM
Tlg. HILAL AHMAR ABU DHABI

British Red Cross

9 Grosvenor Crescent
London SW1X 7EJ
UNITED KINGDOM
Tel. (44)(171) 2355454
Fax (44)(171) 2456315
Tlx 918657 BRCS G
Tlg. REDCROS, LONDON, SW1
E-mail
lorna_finnegan@brcsnhq.org

American Red Cross

17th and D Streets NW
Washington, DC 20006
UNITED STATES
Tel. (1)(202) 7286600
Fax (1)(202) 7286404
Tlx ARC TLX WSH 892636
Tlg. AMCROSS
WASHINGTON DC
E-mail
postmaster@usa.redcross.org

Uruguayan Red Cross

Avenida 8 de Octubre, 2990
11600 Montevideo
URUGUAY
Tel. (598)(2) 802112
Fax (598)(2) 800714
Tlg. CRUZ ROJA
URUGUAYA MONTEVIDEO

Red Crescent Society of Uzbekistan

30, Yusuf Hos Hojib St.
700031 Tashkent
UZBEKISTAN
Tel. (7)(3712) 563741
Fax (7)(3712) 561801

**Vanuatu Red Cross
Society**

P.O. Box 618
Port Vila
VANUATU
Tel. (678) 22599
Fax (678) 22599
Tlx   VANRED
Tlg. VANRED

Venezuelan Red Cross

Avenida Andrés Bello, 4
Caracas 1010
Postal Address:
Apartado 3185
Caracas 1010
VENEZUELA
Tel. (58)(2) 5714380;
5712143; 5715435;
5715957
Fax (58)(2) 5761042;
5715435
Tlx 27237 CRURO VC
Tlg. CRUZ ROJA CARA-
CAS

Red Cross of Viet Nam

68, Rue Ba Triêu
Hanoï
VIET NAM
Tel. (844)(8) 262315;
264868
Fax (844)(8) 266285
Tlx 411415 VNRC VT
Tlg. VIETNAMCROSS
HANOI
E-mail
vnrchq@netnam.org.vn

**Yemen Red Crescent
Society**

Head Office, Building N 10
26 September Street
Sanaa
Postal Address:
P.O. Box 1257
Sanaa
 YEMEN
Tel. (967)(1) 283132;
283133
Fax (967)(1) 283131
Tlx 3124 HILAL YE
Tlg. SANAA HELAL AH-
MAR

Yugoslav Red Cross

Simina 19
11000 Belgrade
YUGOSLAVIA
Tel. (381)(11) 623564
Fax (381)(11) 622965
Tlx 11587 YU CROSS
Tlg. YUGOCROSS
BELGRADE

**Red Cross Society of the
Republic of Zaïre**

41, Avenue de la Justice
Zone de la Gombe
Kinshasa I
Postal Address: B.P. 1712
Kinshasa I
ZAIRE
Tel. (243)(12) 34897
Tlx 21301
Tlg. ZAIRECROIX KIN-
SHASA (BP 1712)

**Zambia Red Cross
Society**

2837 Los Angeles Boulevard
Longacres
Lusaka
Postal Address:
P.O. Box 50001
 (Ridgeway 15101)
Lusaka
ZAMBIA
Tel. (260)(1) 250607;
254798
Fax (260)(1) 252219
Tlx ZACROS ZA 45020
Tlg. REDRAID LUSAKA
E-mail zrcs@zamnet.zm

**Zimbabwe Red Cross
Society**

Red Cross House
98 Cameron Street
Harare
Postal Address:
P.O. Box 1406
Harare
ZIMBABWE
Tel. (263)(4) 775416;
773912
Fax (263)(4) 751739
Tlx 24626 ZRCS ZW
Tlg. ZIMCROSS HARARE

Supporting national Red Cross and Red Crescent societies, mobilizing resources, coordinating action; the global network of International Federation delegations helps to realize the potential of the largest grassroots organization in the world.
Bosnia, 1996.

Chapter

13 The International Federation network of regional and country delegations

Contact details for regional and country delegations of the International Federation of Red Cross and Red Crescent Societies.* Please forward any corrections to the International Federation's Information Resource Centre in Geneva (e-mail: irc@ifrc.org).

The International Federation of Red Cross and Red Crescent Societies

P.O. Box 372, 1211 Geneva 19
SWITZERLAND
Tel. (41)(22) 730 42 22
Fax (41)(22) 733 03 95
Tlx (045) 412 133 FRC CH
Tlg. LICROSS GENEVA
E-mail secretariat@ifrc.org
WWW http://www.ifrc.org

* Information correct as of 1 March 1997.

Red Cross/EU Liaison Bureau

Rue J. Stallaert 1, bte 14
1050 - Bruxelles, BELGIUM
Tel. (32)(2) 3475750
Fax (32)(2) 3474365

International Federation of Red Cross and Red Crescent Societies at the United Nations

630 Third Avenue, 21st floor, Suite 2104,
New York, NY10017, UNITED STATES
Tel. (1)(212) 3380161; Fax (1)(212) 3389832

International Federation regional delegations

Buenos Aires

Lucio V. Mansilla 2698 2o
1425 Buenos Aires
ARGENTINA
Tel. (54)(1) 9638659;
9638690
Fax (54)(1) 9613360

Sydney

Suite 3, Level 3, "Carlton
House"
36-38 York Street
Sydney NSW 2000
AUSTRALIA
Tel. (61)(2) 92991355
Fax (61)(2) 92992320

Brazzaville

Rue des Compagnons de
Brazza
BP 88
Brazzaville
CONGO
Tel. (242) 834252
Fax (242) 833933

San José

Apartado 7-3320
San José 1000
COSTA RICA
Tel. (506) 2326565;
2327575
Fax (506) 2328383

Abidjan

B.P. 2090
Abidjan 04
COTE D'IVOIRE
Tel. (225) 212891; 224853
Fax (225) 212562
Tlx (0983) 22673 LRCS CI

Budapest

Zolyomi Lepcso Ut 22
1124 Budapest
HUNGARY
Tel. (36)(1) 3193423;
3193425
Fax (36)(1) 3193424

Kingston

12 Upper Montrose Road
Kingston 6
Postal Address:
P.O. Box 1284
Kingston 8
JAMAICA
Tel. (1)(809) 9277971
Fax (1)(809) 9789950

Amman

Al Shmeisani
Maroof Al Rasafi Street
Building No. 19
Amman
Postal Address:
P.O. Box 830511 / Zahran
Amman
JORDAN
Tel. (962)(6) 681060
Fax (962)(6) 694556

Almaty

c/o Red Cross and Red
Crescent Society of
Kazakhstan
86, Kunaeva Street
480100 Almaty
KAZAKHSTAN
Tel. (7)(3272) 542742;
542743
Fax (7)(3272) 541535
 Tlx (064) 212378 ifrc su

Nairobi

Chaka Road (off Argwings
Kodhele)
P.O. Box 41275
Nairobi
KENYA
Tel. (254)(2) 714255;
714256
Fax (254)(2) 718415
Tlx (0987) 22622 IFRC KE

Kuala Lumpur

c/o Malaysian Red
Crescent Society
32, Jalan Nipah
Off Jalan Ampang
55000 Kuala Lumpur
MALAYSIA
Tel. (60)(3) 4510723;
4524046
Fax (60)(3) 4519359

South Asia

c/o Nepal Red Cross
Society
Red Cross Marg, Kalimati
Kathmandu
NEPAL
Tel. (977)(1) 273747
Fax (977)(1) 273747

International Federation country delegations

Afghanistan

43 D Jamal-ud-Din
Afghan Road
University Town
Peshawar
PAKISTAN
Tel. (8731) 754374
Fax (8731) 754374

Angola

Rua Emilio M'Bidi 51 - 51A
Bairro Alvalade
Luanda
Postal Address:
Caixa Postal 3324
 Luanda
ANGOLA
Tel. (244)(2) 324448
Fax (244)(2) 320648
Tlx 3394 crzver an

Armenia

Djrashati Street 96
Yerevan-19
ARMENIA
Tel. (3742) 522253;
561889

Moscow

c/o Russian Red Cross
Tcheremushkinski
Proezd 5
117036 Moscow
RUSSIAN FEDERATION
Tel. (7)(095) 2306620;
2306621
Fax (7)(095) 2306622

Harare

11, Phillips Avenue
Belgravia
Harare
ZIMBABWE
Tel. (263)(4) 720315;
720316
Fax (263)(4) 708784
Tlx (0907) 24792

Azerbaijan

Niazi Street 11
Baku 370000
AZERBAIJAN
Tel. (99)(412) 925792
Fax (99)(412) 931889
Tlx (064) 142454

Bangladesh

c/o Bangladesh Red
Crescent Society
684-686 Bara Magh Bazar
Dhaka - 1217
BANGLADESH
Tel. (880)(2) 835148
Fax (880)(2) 834701

Belarus

Ulitsa Mayakovkosgo 14
Minsk 220 006
BELARUS
Tel. (375)(172) 217237;
219060
Fax (375)(172) 219060

Benin

B.P. 08-1070
Lot A9 - Les Cocotiers
Haie Vive, Cotonou
BENIN
Tel. (229) 301013
Fax (229) 300614

Bosnia and Herzegovina

Titova 7
71000 Sarajevo
BOSNIA AND
HERZEGOVINA
Tel. (387)(71) 666009;
666010
Fax (387)(71) 666011

Burundi

Avenue des Etats-Unis
3674A
Bujumbura
Postal Address: B.P. 324
Bujumbura
BURUNDI
Tel. (257) 216246; 211101
Fax (257) 229408

Cambodia

53 Deo, Street Croix-Rouge
Phnom Penh
 Postal Address: Central
Post Office/P.O. Box 620
Phnom Penh
CAMBODIA
Tel. (855)(23) 362690;
426370
Fax (855)(23) 426599

Congo

c/o Comité de la
Croix-Rouge congolaise
B.P. 650
Pointe-Noire
CONGO
Tel. (242) 948425
Fax (242) 945471

Côte d'Ivoire

c/o Croix-Rouge de Côte
d'Ivoire
B.P. 1244
Abidjan 01
COTE D'IVOIRE
Tel. (225) 321529
Fax (225) 225355

Croatia

Florijana Andreaseca 14
10000 Zagreb
CROATIA
Tel. (385)(1) 306000
Fax (385)(1) 334251

Eritrea

Quohaito Street No 1-3
Asmara
Postal Address: c/o Red
Cross Society of Eritrea
P.O. Box 575, Asmara
ERITREA
Tel. (291)(1) 127857
Fax (291)(1) 124198

Ethiopia

Ras Destra Damtew
Avenue
Addis Ababa
Postal Address: c/o
Ethiopian Red Cross
P.O. Box 195
Addis Ababa
ETHIOPIA
Tel. (251)(1) 514571
Fax (251)(1) 512888
Tlx 21338 ERCS

Georgia

7, Anton Katalikosi Street
Tbilisi
GEORGIA
Tel. (995)(32) 988365
Fax (995)(32) 985976

Ghana

Ring Road Central House
No C 263/3, 1st floor
next to Provident
Insurance Tower
Accra
 GHANA
Tel. (233)(21) 232133
Fax (233)(21) 232133

Guinea

c/o Croix-Rouge de
Guinée
B.P. No 376
Conakry
GUINEA
Tel. (224) 413825
Fax (224) 414255

Haïti

BP 15322
Pétionville
HAITI
Tel. (509) 570445; 578818
Fax (509) 578819

India

2-C2 Parkwood Apts.
Rao Tula Ram Marg
(opp. Shanti Niketan)
New Delhi -110 022
INDIA
Tel. (91)(11) 6164760;
6164762
Fax (91)(11) 6166661

Kenya

Links Road, Off Mombasa
Malindi Road (Nr Reef Hotel)
Mombasa
Postal Address:
P.O. Box 34099
Mombasa
KENYA
Tel. (254)(11) 486487;
486488
Fax (254)(11) 471345
Tlx 22622 ifrc ke

Democratic People's Republic of Korea

c/o Red Cross Society of
the DPR Korea
Ryonwa 1, Central District
Pyongyang
KOREA, DEMOCRATIC
PEOPLE'S REPUBLIC OF
Tel. (850)(2) 18111; 18222
Fax (850)(2) 3814644

Lao People's Democratic Republic

Setthatirath Road
Xiengnhune Vientiane
Postal Address: c/o Lao
Red Cross
P.O. Box 2948
Vientiane
LAO PEOPLE'S
DEMOCRATIC REPUBLIC
Tel. (856)(21) 215762;
215935
Fax (856)(21) 215935
Tlx 4491 TE VTE LS

Lebanon

N. Dagher Building
 Mar Tacla - Beirut
LEBANON
Tel. (961)(1) 424851
Fax (1)(212) 4448553
(New York)

Lesotho

c/o Lesotho Red Cross
23 Mabile Road
Maseru 100
LESOTHO
Tel. (266) 313911
Fax (266) 317984
Tlx 4515 LO

Liberia

107, Lynch Street
Monrovia
Postal Address: c/o
Liberian Red Cross Society
P.O. Box 5081
Monrovia
LIBERIA
Tel. (231) 227485; 226231
Fax (231) 226231

Madagascar

c/o Malagasy Red Cross
Society
1, rue Patrice Lumumba
Tsavalalana
Antananarivo
Postal Address: B.P. 1168
Antananarivo
MADAGASCAR
Tel. (261)(2) 22111
Fax (261)(2) 35457

Malawi

Red Cross House
Lilongwe 3
Postal Address: c/o
Malawi Red Cross
P.O. Box 30096
Lilongwe 3
MALAWI
Tel. (265) 732877; 732878
Fax (265) 731403
Tlx (0904) 44276

Mali

c/o Croix-Rouge malienne
B.P. 151
Gao
MALI
Tel. (223) 820190
Fax (223) 820190

Mongolia

c/o Red Cross Society of
Mongolia
Central Post Office
P.O. Box 537
Ulaan Baatar
MONGOLIA
Tel. (976)(1) 321684
 Fax (976)(1) 321684
Tlx 79358 MUZN

Mozambique

Avenida 24 de Julho, 641
Maputo
Postal Address:
Caixa postal 2488
Maputo
MOZAMBIQUE
Tel. (258)(1) 421210
Fax (258)(1) 423507
Tlx (0992) 6-169 CV Mo

Myanmar

c/o Myanmar Red Cross
Society
Red Cross Building
42 Strand Road
Yangon
MYANMAR
Tel. (95)(1) 297877
Fax (95)(1) 297877
Tlx 21218

Namibia

100, Robert Mugabe
Avenue
Windhoek
Postal Address: c/o
Namibia Red Cross
Society
P.O. Box 346
Windhoek
NAMIBIA
Tel. (264)(61) 222135
Fax (264)(61) 228949

Nepal

c/o Nepal Red Cross
Society
Red Cross Marg, Kalimati
Kathmandu
Postal Address:
P.O. Box 217
Kathmandu
NEPAL
Tel. (977)(1) 272761;
270650
Fax (977)(1) 273747
Tlx 2569 NRCS NP

Nigeria

11, Eko Akete Close
Off St. Gregory's Road
South West Ikoyi
Lagos
Postal Address: c/o
Nigerian Red Cross
Society
P.O. Box 764
Lagos
NIGERIA
Tel. (234)(1) 2695188;
2695189
Fax (234)(1) 680865

Papua New Guinea

c/o Papua New Guinea
Red Cross Society
Taurama Road,
Port Moresby, Boroko
Postal Address:
P.O. Box 6545
Boroko
PAPUA NEW GUINEA
Tel. (675) 3258577;
3258759
Fax (675) 3259714
Tlx PNG RC NE 23292

Rwanda

Plot No 317
Kimihurura - Kacyiru
Kigali
RWANDA
Tel. (250) 74402; 750088
Fax (250) 750088

Sierra Leone

6, Liverpool Street
Freetown
Postal Address: c/o Sierra
Leone Red Cross Society
P.O. Box 427
Freetown
SIERRA LEONE
Tel. (232)(22) 228180;
222384
Fax (232)(22) 228180
Tlx 3692 SLRCS SL

Somalia

Lenana Road
Kilimani, Nairobi
Postal Address: P.O. Box
41275, Nairobi
KENYA
Tel. (254)(2) 564602;
564623
Fax (254)(2) 564750
Tlx 25436 IFRC KE

Sri Lanka

120 Park Road
Colombo 5
SRI LANKA
Tel. (94)(1) 592159;
581903
Fax (94)(1) 583269

Sudan

Al Mak Nimir
Street/Gamhouria Street
Plot No 1, Block No 4
East Khartoum
Postal Address:
P.O. Box 10697
East Khartoum
SUDAN
Tel. (249)(11) 770484
 Fax (249)(11) 770484
Tlx (0984) 23006 LRCS SD

Tanzania

Ali Hassan Mwinyi
Dar es Salaam
Postal Address:
P.O. Box 1133,
Dar es Salaam
TANZANIA, UNITED
REPUBLIC OF
Tel. (255)(51) 116514
Fax (255)(51) 116514

Turkey

Atatürk Bulvari
219/14 Bulvari Apt.
006680 Kavaklidere,
Ankara
TURKEY
Tel. (90)(312) 4672099;
4673349
Fax (90)(312) 4274217

Uganda

Plot 97, Buganda Road
Kampala
Postal Address: c/o
Uganda Red Cross
Society
P.O. Box 494
Kampala
UGANDA
Tel. (256)(41) 231480;
243742
Fax (256)(41) 258184

Viet Nam

19 Mai Hac De Street
Hanoï
VIET NAM
Tel. (84)(4) 8252250;
8229283
Fax (84)(4) 8266177
Tlx (0805) 411415 VNRC
VT

Yugoslavia

Simina Ulica Broj 21
11000 Belgrade
YUGOSLAVIA
Tel. (381)(11) 3282202;
3281376
Fax (381)(11) 3281791

Zaire

Concession Boukin, 4
Z/Ngaliema, Kin 1
Kinshasa
ZAIRE
Tel. (243)(88) 45582;
40182

The International Federation of Red Cross and
Red Crescent Societies

1996 Relief Operations

Maps

In 1996, appeals for over 320 million Swiss francs were launched to assist more than 15 million people.

The maps on the following pages summarize the following operations:

- natural disasters, including floods, hurricanes and cyclones and earthquakes;
- population movements;
- socio-economic displacements;
- others, including food shortages, epidemics and nuclear disaster; and
- regional programmes.

Floods

Hurricanes
Cyclones

Earthquakes

Floods

Country/ Programme (Date)	Number of victims assisted	Aid sought in cash/kind/ services (Swiss Francs)
Morocco (26/01)	18,000	1,265,000
DPR Korea (22/03)	130,000	7,658,000
China (08/07)	1,470,000	5,442,000
Bangladesh (05/08)	100,000	650,000
Viet Nam (21/08)	90,000	747,000
Viet Nam (17/10)	386,500	1,501,000
Honduras (03/12)	12,000	240,000

Hurricanes/Cyclones

Central America (02/08)	21,175	1,134,000
Cuba (21/10)	80,000	2,126,000
India (08/11)		1,000,000

Earthquakes

Ecuador (04/04)	3,000	467,000
Peru (15/11)	10,000	913,000
TOTAL	**2,320,675**	**23,143,000**

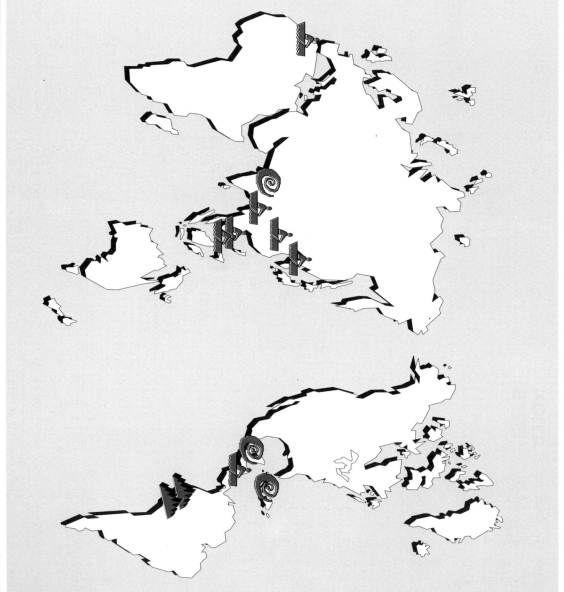

POPULATION MOVEMENTS

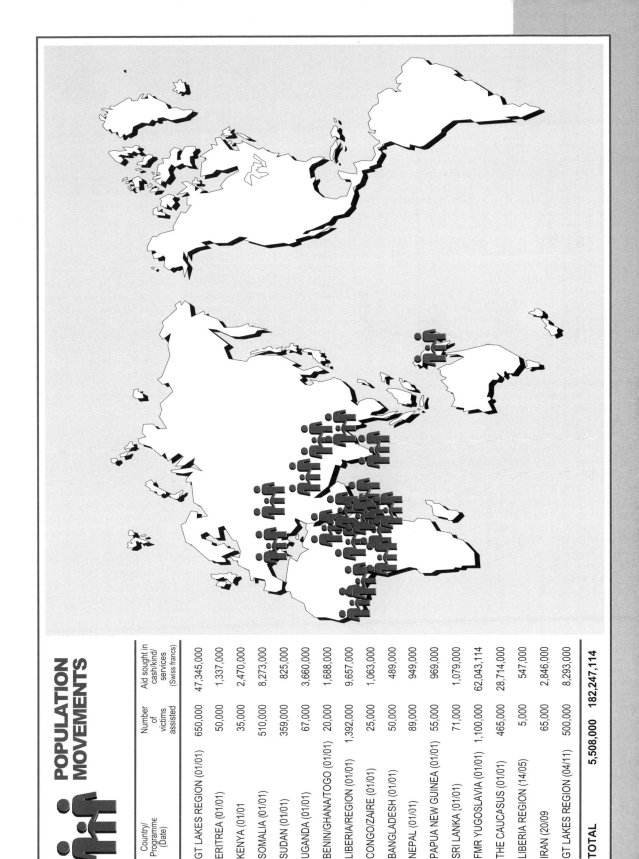

Country/ Programme (Date)	Number of victims assisted	Aid sought in cash/kind/ services (Swiss francs)
GT LAKES REGION (01/01)	650,000	47,345,000
ERITREA (01/01)	50,000	1,337,000
KENYA (01/01)	35,000	2,470,000
SOMALIA (01/01)	510,000	8,273,000
SUDAN (01/01)	359,000	825,000
UGANDA (01/01)	67,000	3,660,000
BENIN/GHANA/TOGO (01/01)	20,000	1,688,000
LIBERIA/REGION (01/01)	1,392,000	9,657,000
CONGO/ZAIRE (01/01)	25,000	1,063,000
BANGLADESH (01/01)	50,000	489,000
NEPAL (01/01)	89,000	949,000
PAPUA NEW GUINEA (01/01)	55,000	969,000
SRI LANKA (01/01)	71,000	1,079,000
FMR YUGOSLAVIA (01/01)	1,100,000	62,043,114
THE CAUCASUS (01/01)	465,000	28,714,000
LIBERIA REGION (14/05)	5,000	547,000
IRAN (20/09)	65,000	2,846,000
GT LAKES REGION (04/11)	500,000	8,293,000
TOTAL	**5,508,000**	**182,247,114**

Country/ Programme (Date)	Number of victims/ assisted	Aid sought in cash/kind/ services (Swiss francs)
HAITI (01/01)	160,000	3,677,000
AFGHANISTAN (01/01)	1,500,000	6,310,000
CAMBODIA (01/01)	221,500	2,845,000
TAJIKISTAN/ KYRGYZSTAN (01/01)	1,128,000	16,724,000
RUSSIAN FEDERATION (01/01)	601,000	12,519,000
IRAQ (01/01)	300,000	17,520,000
PALESTINE RED CRESCENT SOCIETY (01/01)		10,737,000
ANGOLA (15/03)	600,000	7,807,000
BULGARIA (22/10)	24,500	659,000
TOTAL	**4,535,000**	**78,798,000**

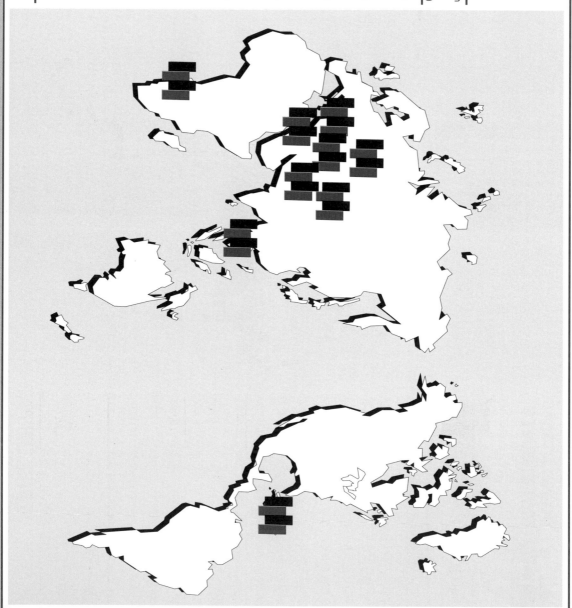

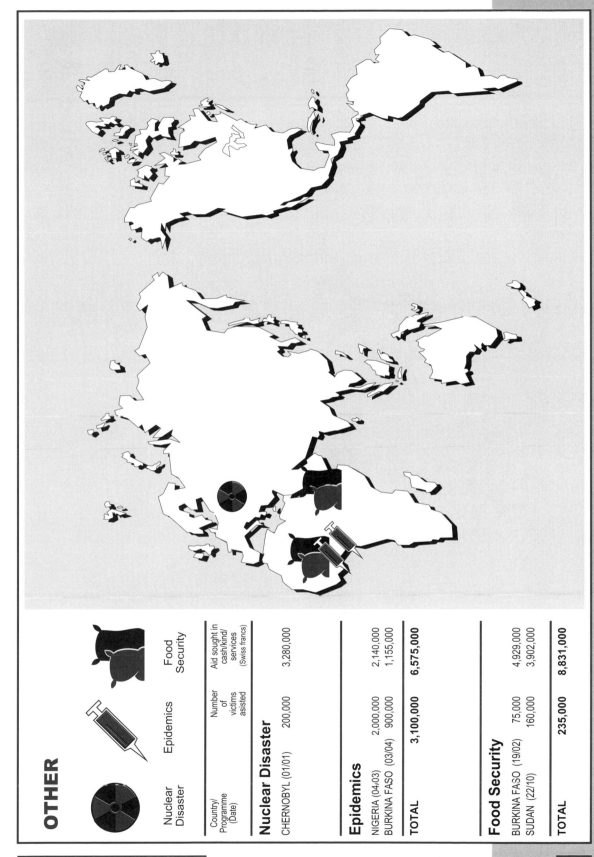

OTHER

	Nuclear Disaster	Epidemics	Food Security

Country/ Programme (Date)	Number of victims asisted	Aid sought in cash/kind/ services (Swiss francs)

Nuclear Disaster

CHERNOBYL (01/01)	200,000	3,280,000

Epidemics

NIGERIA (04/03)	2,000,000	2,140,000
BURKINA FASO (03/04)	900,000	1,155,000
TOTAL	**3,100,000**	**6,575,000**

Food Security

BURKINA FASO (19/02)	75,000	4,929,000
SUDAN (22/10)	160,000	3,902,000
TOTAL	**235,000**	**8,831,000**

REGIONAL PROGRAMMES

	Aid sought in cash/kind services (Swiss francs)
West Africa	2,630,000
Central Africa	6,229,000
East Africa	1,674,000
Southern Africa	1,955,000
Central and Eastern Europe	1,839,000
Minsk Delegation	1,074,000
Amman Delegation	868,000
Central Asia	2,781,000
South Asia	433,000
South and East Asia	2,701,000
Pacific	1,980,000
Caribbean	729,000
Central and South America	650,000

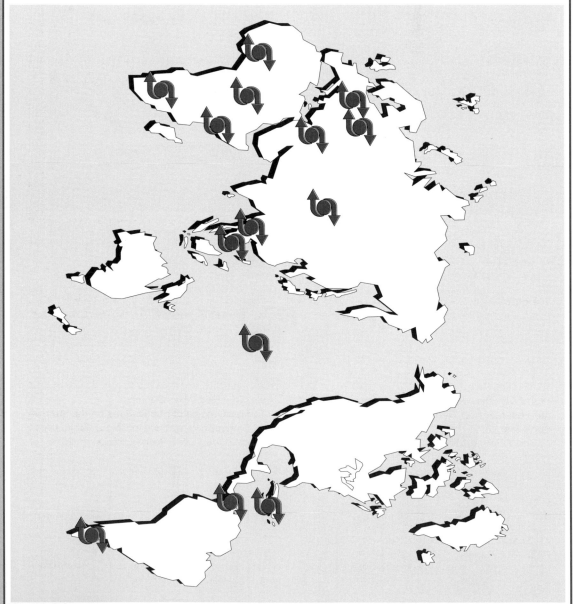

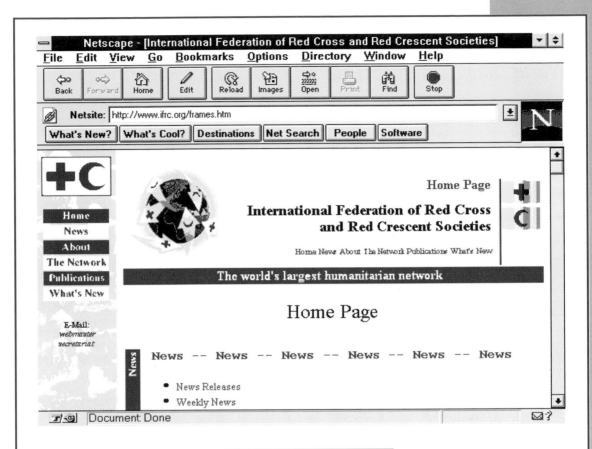

The International Federation on the Internet

Internet users can access a wide range of information, including the full text of the *World Disasters Report*, from the International Federation of Red Cross and Red Crescent Societies.

Internet users can access a wide range of information, including selected chapters from previous *World Disasters Reports* and the full text of the *World Disasters Report 1997*, from the International Federation of Red Cross and Red Crescent Societies.

The International Federation pages on the World-Wide Web include constantly-updated disaster appeals and situation reports from its global network of operations and delegations, as well as links to many disaster-related sites and the web pages of the following:

International Committee of the Red Cross, International Red Cross and Red Crescent Museum, Henry Dunant Institute, American Red Cross, Austrian Red Cross, Belgian Red Cross, The Canadian Red Cross Society, Red Cross Society of China,

Colombian Red Cross Society, Croatian Red Cross, Danish Red Cross, Finnish Red Cross, German Red Cross, Irish Red Cross Society, Italian Red Cross, Japanese Red Cross Society, Mexican Red Cross, The Netherlands Red Cross, Norwegian Red Cross, Polish Red Cross, Singapore Red Cross Society, Spanish Red Cross, Swedish Red Cross, Thai Red Cross Society.

To make copies of publications on the International Federation's web site or to publish extracts of them, contact the Secretariat in advance for permission (Secretariat e-mail address: secretariat@ifrc.org). Full acknowledgement will be required.

The International Federation pages are at

http://www.ifrc.org

If you have any problems connecting or would like more information, please contact the International Federation's Internet manager, Jeremy Mortimer, on webmaster@ifrc.org

The world of disasters in the

World Disasters Report 1993-1997

The *World Disasters Report* is the only annual, interdisciplinary report focusing on disasters, from natural hazards to human-induced crises, and the millions of people affected by them.

Published in English and a range of other languages each year since 1993, the *World Disasters Report* analyses cutting-edge issues, assesses practical methodologies, examines recent experience and collates a comprehensive disasters database. It is backed by the expertise and resources of the International Federation of Red Cross and Red Crescent Societies, whose relief operations and National Societies in 170 countries make it the most extensive and experienced humanitarian network.

To order further copies of the *World Disasters Report* 1997 or the 1996 edition in English, use the order form on page 174; for all other languages and years contact the International Federation.

The all-new *World Disasters Report 1997* examines the military-humanitarian relationship, information management and epidemiological data collection in disasters, aid trends and standards in disaster response, devastating floods in China, natural hazards in the Caribbean, the challenge of Somalia, and re-emerging diseases in the former Soviet Union. The Report includes a comprehensive 25-year disasters database, and outlines the global activities of the International Federation. Indexed and fully illustrated. Maps.

The five-section *World Disasters Report 1996* includes: global population movements, causes and consequences; global food security, including issues of gender, conflict and rights; emergency food aid and nutrition; developmental relief; trends in aid; surviving the Kobe earthquake; challenges of Rwanda; Oklahoma's trauma; DPR Korea's food and flood crisis; meeting the need for systematic data; Code of Conduct update; full listings of National Societies and delegations; 25-year disasters database. Indexed and fully illustrated.

The four-section *World Disasters Report 1995* includes: UN sanctions and the humanitarian crisis, with case histories from Iraq, Serbia, Haiti; good disaster-relief practice; turning early warning into livelihood monitoring; measuring the effects of evaluation; listening to the beneficiaries; psychological support; humanitarians in uniform; hate ratio in conflict; mines and demobilization; surviving cyclones in Bangladesh; Ethiopia ten years on; success and failure in Rwanda; working in Somalia's grey zone. Fully illustrated.

The three-section *World Disasters Report 1994* includes: drought success in Southern Africa; conflict and progress in Somalia; agency challenges within the former Yugoslavia; Brazil's vulnerability; India's earthquake myths; Caucasus collapse; secrecy's role in disasters; global survey of anti-personnel mines, information and Chernobyl; African peace mechanisms; human rights and disasters; indigenous knowledge and response; and the full text of the Code of Conduct for disaster-relief agencies.

The *World Disasters Report 1993* the pilot issue includes: humanitarian gap, preparedness versus relief, role of foreign medical teams and military forces, equity in impact, media in disasters, AIDS, famine, flood, high winds, refugees, epidemics, earthquakes, volcanoes. Case histories from: Uganda, Sudan, China, Bangladesh, Afghanistan, Peru, Zambia, Turkey, United States, Philippines. Fully illustrated.

And coming up ...

World Disasters Report 1998

The 1998 edition of the *World Disasters Report* will examine increasing urbanization and population growth, consider the problems of megacities and their shanty towns, ageing populations, inmigration and the poverty-push, hunger and disease, risk and hazards – how will all disaster stakeholders cope and what will it mean for disaster preparedness, mitigation and response?

The Report will also look at the year in disasters 1997, and include its disaster database, Code of Conduct update, and full listings of National Societies and delegations.

Index

How to order the *WORLD DISASTERS REPORT 1997* from Oxford University Press*

CREDIT CARD HOTLINE
Phone OUP's Credit Card Hotline, open 24 hours a day, quoting the reference number in the bottom right corner:
Tel.: +44 (0) 1536 454 534

FAX ORDERS
Fax: +44 (0) 1536 746 337

TELEPHONE ENQUIRIES
Tel: +44 (0) 1536 741 519

BY E-MAIL
E-mail: orders@oup.co.uk

BY POST
From the UK (no stamp required):
CWO Department, OUP, FREEPOST, NH4051, Corby, Northants, NN18 9BR

From outside the UK:
CWO Department, OUP, Saxon Way West, Corby, Northants, NN18 9ES

HOW TO ORDER IN THE UNITED STATES
For more information, or to order by credit card, please call:

1-800-451-7556

between the hours of 0900 and 1700 Eastern Standard Time

YOUR ORDER

Please supply the following [PLEASE USE BLOCK CAPITALS]:

Qty	ISBN	Author	Title	Price[1]
	0-19-829290-2	IFRC	*World Disasters Report 1997*	£ 15.99
				£
			EC customers from outside the UK please add VAT[2]	
			Postage and Packing[3]	
			TOTAL	

DELIVERY

Please deliver my goods to:[4]

Title	
First Name	
Last Name	
Department/Faculty	
University/Company	
Address	
Country	
Post Code	
E-mail Address	

HOW TO PAY

You can pay by: ☐ Credit Card ☐ Cheque from a UK Bank account ☐ Eurocheque
PLEASE COMPLETE THE RELEVANT FORM BELOW:

Please charge £_____ to my MasterCard / Visa / American Express / Diners Club account

Card Number ☐☐☐☐ ☐☐☐☐ ☐☐☐☐ ☐☐☐☐ ☐☐☐☐

Expiry Date ____/____ Signature _____

Credit Card Account Address (if different from Delivery Address):
_____
_____
_____ I enclose a Eurocheque for £_____

☐ **CHEQUE PAYMENT**
I enclose a cheque for £_____
crossed and made payable to **Oxford University Press**, drawn against a UK Bank.

☐ **EUROCHEQUE PAYMENT**

☐ **PROFORMA INVOICE**
Please send a proforma invoice for £_____
Goods will be despatched on receipt of payment.

Please quote the following code for all telephone or e-mail orders: **ANREDX97 Z**

[1] Prices and Extents are accurate at the time of going to press, but are liable to alteration without notice.

[2] If you are registered for VAT or a local sales tax, please provide your number here:

[3] Postage and Packing charges (including VAT)

UK orders
under £20, add £2.06
over £20, add £3.53
over £50, add £4.70

Non-UK orders
add 10 per cent of the total price of the goods

[4] Please allow 10 days for delivery in the UK; 28 days elsewhere.

☐ Tick here if you do not want to be sent information about OUP titles in the future.

Thank you for your order!

* The *World Disasters Report 1996* in English can also be ordered through OUP.

For orders and information about 1993, 1994 and 1995 editions in English and all other language versions, please contact:

International Federation of Red Cross and Red Crescent Societies
P.O. Box 372, 1211 Geneva 19, Switzerland

Tel: +41 22 730 4222;
fax: +41 22 733 0395;
e-mail: guidera@ifrc.org

A world of humanitarian news in one magazine

Red Cross Red Crescent is the dynamic full-colour magazine that uses powerful writing and dramatic pictures to cover the global work of the International Red Cross and Red Crescent Movement and issues of concern to the entire humanitarian community.

Published three times a year, in English, French and Spanish, *Red Cross Red Crescent* offers front line news of conflict and disaster, assesses current crises, highlights important campaigns and covers the views of people dealing with humanitarian issues.

Recent and upcoming topics in *Red Cross Red Crescent* include: the campaign to ban antipersonnel mines, aid and media, challenges facing Colombia, Russia in transition, Asia's growing cities, AIDS and human rights, blinding laser weapons, TV and policy making, NGO independence.

Red Cross Red Crescent is produced jointly by the International Committee of the Red Cross and the International Federation of Red Cross and Red Crescent Societies.

For more information, or to request a free subscription to *Red Cross Red Crescent*, please write with your details – title, name, full postal address, phone, fax and e-mail – plus indications of your professional or personal areas of interest, and the language you would prefer, to:

**The Editors,
Red Cross Red Crescent,
PO Box 372, CH 1211 Geneva 19,
Switzerland
Fax: +41 22 733 0395.
E-mail: grisewood@icrc.org**

Members of national Red Cross and Red Crescent societies: please contact your branch or headquarters

---

From the International Federation of Red Cross and Red Crescent Societies, the

Annual Report 1997

reviews the International Federation's work through 170 national Red Cross and Red Crescent societies, worldwide delegations and its Geneva Secretariat. The Report examines humanitarian action; capacity building; communications; external relations; revenue generation; governance, management and accountability.

To request a free copy in English, French, Spanish or Arabic, send your full details to:

**Publications Department,
International Federation of Red Cross
and Red Crescent Societies,
PO Box 372, 1211 Geneva 19, Switzerland;
Fax: +41 22 733 0395; E-mail: guidera@ifrc.org**

Play PLUS Lotto

PLUS Lotto, the world's first global Internet lottery supporting a humanitarian cause, gives you the chance to win great prizes – and 25% of all proceeds are donated to the International Federation and National Societies to help deliver services when and where they are needed most.

PLUS Lotto is operated by the charitable InterLotto Foundation, located in Liechtenstein. Every time you play, someone, somewhere in the world or in your country is a winner!

Play it any time of the day or night at:
http://www.pluslotto.com

For more information visit these web sites:

International Committee of the Red Cross **http://www.icrc.org**
International Federation of Red Cross and Red Crescent Societies **http://www.ifrc.org**

OXFORD JOURNALS

African Affairs

African Affairs is one of the most presitious journals in the field. It publishes articles on recent political, social and economic developments in sub-Saharan countries, and includes historical studies that illuminate current events in the continent.
Volume 96, 1997 (4 issues)
Institutions: £70/US$125, Individuals: £39/US$64, Individuals in Africa: US$28

African Languages and Cultures

African Languages and Cultures publishes articles dealing with the descriptive linguistics of African languages, language classification and comparative African linguistics, African cultural studies, African language literatures, African writing in metropolitan languages, African art and African music.
Volume 10, 1997 (2 issues)
Institutions: £32/US$54, Individuals: £18/US$32, Developing countries: US$32

Journal of African Economies

In the last few years there has been a growing output of high quality economic research on Africa, but until the advent of the **Journal of African Economies** it was scattered over many diverse publications. Now this important area of research has its own vehicle to carry rigorous economic analysis, focused entirely on Africa, for Africans and anyone interested in the continent.
Volume 6, 1997 (3 issues)
Institutions: £75/US$128, Individuals: £41/US$76, All Subscribers in Africa: US$50

Journal of African Law

The **Journal of African Law** is essential reading for those interested in contemporary legal developments in Africa. The journal covers the full range of legal fields, including crime, family law, commercial law, human rights, and nationality and constitutional issues. It also contains notes on recent legislation, cases, reform proposals, and international developments.
Volume 41, 1997 (2 issues)
Institutions: £53/US$85, Individuals: £30/US$50, Developing countries: US$50

Bulletin of SOAS

The **Bulletin of the School of Oriental and African Studies** spans the cultures and civilizations, from ancient times to the present day, of the Near and Middle East, South Asia, Central Asia, East Asia, South-East Asia, and Africa. It publishes articles and shorter notes on history, religions, philosophies, literatures, languages, music, art, archaeology, law, and anthropology.
Volume 60, 1997 (3 issues)
Institutions: £76/US$134, Individuals: £49/US$79, Institutions in developing countries: US$110

International Journal of Refugee Law

A journal which aims to stimulate research and thinking on refugee law and its development, taking account of the broadest range of State and international organisation practice. It serves as an essential tool for all engaged in the protection of refugees and finding solutions to their problems, providing key information and commentary on today's critical issues.
Volume 9, 1997 (4 issues), Institutions: £78/US$145, Individuals: £38/US$71

Refugee Survey Quarterly

RSQ is produced by the Centre for Documentation and Research of the United Nations High Commissioner for Refugees. Published four times a year it serves as an authoritative source for current refugee and country information. Each issue is a combination of country reports, documents, reviews and abstracts of refugee-related literature.
Volume 16, 1997 (4 issues), Institutions: £62/US$92, Individuals: £41/US$65

Journal of Refugee Studies

The **Journal of Refugee Studies** provides a major focus for research into refugees reflecting the diverse range of issues involved. It aims to promote the theoretical development of refugee studies, and encourages the voice of refugees.
Volume 10, 1997 (4 issues), Institutions: £69/US$127, Individuals: £35/US$62
Institutions in developing countries: US$57

The Adelphi Papers

The **Adelphi Paper** series offers rigorous analysis of strategic and defence topics and other important issues of the day. The 1997 **Adelphi Papers** have been redesigned to offer an increased number of maps, charts and graphs for easy reference.
8-10 Papers, 1997, Subscription rate: £104/US$168

Survival

Survival, The IISS Quarterly, is essential reading for practitioners, analysts, teachers and followers of international affairs and security. Topics covered include NATO, the UN, European security and much more.
Volume 39, 1997 (4 issues), Institutions: £43/US$68, Student rate: £25/US$39

OXFORD
JOURNALS

The Economics of Transition

The Economics of Transition provides a reliable and authoritative analysis of the transition process in the former communist states of Europe. It has become the essential forum for reviewing and reporting on the structural issues raised by the process of transition towards market-type economies.
Volume 5, 1997 (2 issues), Institutions: £185/US$300, Academic Libraries: £115/US$185
Individuals: £45/US$74

Health Promotion International

Health Promotion International responds to the move for a new public health awareness throughout the world and supports the development of action outlined in the Ottawa Charter for Health Promotion
Volume 12, 1997 (4 issues), Institutions: £115/US$205, Individuals: £55/US$99

Health Policy and Planning

Particularly relevant to those working in international health planning, medical care and public health, **Health Policy and Planning** is concerned with issues of health policy, planning, and management and evaluation, focusing on the developing world.
Volume 12, 1997 (4 issues)
Institutions: £110/US$195, Individuals: £47/US$88, Developing countries: US$137,
Graduate students: £27/US$49

International Journal of Epidemiology

Exploring the epidemiology of both infectious and non-infectious disease, including research into health services and medical care, **IJE** encourages communication among all those engaged in the research, teaching, and application of epidemiology.
Volume 26, 1997 (6 issues), Subscription rate: £185/US$340, Developing countries: US$204

Community Development Journal

Community Development Journal covers political, economic and social programmes that link the activities of people with institutions and government. Articles feature community action, village, town and regional planning, community studies and rural development.
Volume 32, 1997 (4 issues) Institutions: £54/US$97, Individuals: £39/US$72,
Developing Countries: US$65

The China Quarterly

The China Quarterly is the leading interdisciplinary journal on the People's Republic of China, Taiwan, Macau and Hong Kong, covering politics and international affairs, economics and commerce, geography and demography through to art and literature.
Issues 149-152, 1997 (4 issues) Institutions: £50/US$90, Individuals: £36/US$64,
Students: £18/US$32

Oxford University Press

The State of the World's Refugees, 1997

UNITED NATIONS HIGH COMMISSIONER FOR REFUGEES

During the past few years, millions of people have been forced to abandon their homes, fleeing from communal violence, political persecution and other threats to their security. This invaluable book from the UN's refugee agency provides the definitive picture of the world's displaced people, and examines the international community's efforts to respond to their plight.

0-19-829310-0, 224 pp, numerous photographs, figures, and tables, £30.00
0-19-829309-7, paperback £9.99

Also Available: The State of the World's Refugees, 1995

UNITED NATIONS HIGH COMMISSIONER FOR REFUGEES

0-19-828044-0, 264 pp, numerous halftones, figures and tables, £30.00
0-19-828043-2, paperback £9.99

Human Development Report

1996

UNITED NATIONS DEVELOPMENT PROGRAMME

'Has become among the best of the many publications cataloguing the progress—or lack of it—towards a stable and sustainable society ... This publication both provides a wealth of information and presents it in a form that will stimulate strong debate.' **Global Security**

0-19-511158-3, 240 pp, numerous two-colour illustrations, OUP USA, £25.00
0-19-511159-1, paperback £14.50

Beyond Charity

International Cooperation and the Global Refugee Crisis.
A Twentieth Century Fund Book

GIL LOESCHER

'Beyond Charity is a valuable addition to the literature on refugees and humanitarian assistance, which will be of particular value to the growing number of academics and students now focusing on these issues ... a coherent examination of the global refugee crisis and a balanced assessment of the contribution which UNHCR and other humanitarian agencies can make in their efforts to respond.'
Refugee Survey Quaterly

Too often in the past refugees have been perceived as a problem of international charity. Now, with the end of the Cold War triggering new refugee movements across the globe, Loescher argues that refugee problems are political issues, and must be treated as such. *Beyond Charity* presents a comprehensive overview of refugee problems and assistance programs since World War II, outlining reforms to address both the current refugee crisis and its underlying causes.

0-19-510294-0, 272 pp, line figures, tables, paperback, OUP USA, £13.99

NEW in paperback

DISASTERS

The Journal of Disaster Studies, Policy & Management

Edited by Charlotte Benson and Joanna Macrae, Overseas Development Institute

Disasters is a major peer-reviewed quarterly journal reporting on all aspects of disaster studies, policy and managment. It aims to provide a forum for academics, policy-makers and practitioners on high quality research and practice related to natural disasters and complex political emergencies around the world. The journal promotes the interchange of ideas and experience, and maintains a balance between field reports from relief workers, case studies, articles of general interest and academic papers. It also contains book reviews and conference reports. Letters and discussion are welcomed.

A Disaster Relief Supply Management System

S UMA is a tool for the management of humanitarian supplies, from the time pledges are made by donors, to their entry into the disaster area and their storage and distribution.

◇ **SUMA** keeps authorities and donors informed of exactly what has been received;

◇ **SUMA** quickly identifies and prioritizes those supplies that are urgently needed by the disaster-affected population;

◇ **SUMA** offers a tool for inventory control on warehousing and distribution of supplies; and

◇ **SUMA** can be used in many different emergency situations, large or small scale or in natural as well as complex emergency situations.

The SUMA software is available in English, Spanish, and French. For information on technical requirements and obtaining the software and training manuals, please visit our website or contact:

FUNDESUMA
Apartado Postal 3745 - 1000
San José, Costa Rica

Fax: (506) 257-2139
e-mail: suma@paho.org • fundesuma@netsalud.sa.cr

© Pan American Health Organization

Now you are able to visit SUMA on the internet at: http://www.netsalud.sa.cr/ops/suma/

INCEDE NEWSLETTER

International Center for Disaster-Mitigation Engineering

**Institute of Industrial Science
The University of Tokyo**

An Introduction

INCEDE Newsletter: a herald tribune of disaster information of the decade

INCEDE newsletter, a courier of disaster information, is a quarterly publication of INCEDE and is distributed to INCEDE network and mailing members from 142 countries around the world. The newsletter brings you detailed reports on natural disasters such as typhoons, floods, earthquakes, wild-fires etc. world wide. Also, it contains articles by prominent engineers on general topics related to disaster-mitigation and publishes current activities of INCEDE and its Network Members. In addition to the regular newsletters, INCEDE publishes special newsletter issues covering major scale disasters, such as the 1993 January Kushiro-Oki earthquake, the 1993 Hokkaido-Nansei-Oki earthquake, Hanshin (Kobe) earthquake etc. The newsletters in the past also carried reports on post disaster investigations, conducted either by INCEDE staff or the network members. The INCEDE newsletter Vol. 5, No. 4 which is currently being edited, marks the successful completion of 5 years in print. You can obtain a free copy of the INCEDE Newsletter by joining INCEDE as a network member and it will also bring you the opportunity of sharing information and experiences in disaster mitigation with the large audience of INCEDE network. For more information, browse INCEDE Home page at the following site, or send an e-mail to network@incede.iis.u-tokyo.ac.jp

http://incede.iis.u-tokyo.ac.jp/incede_home.html

At a glance: a few past issues of INCEDE newsletter

INCEDE, Institute of Industrial Science, The University of Tokyo, 7-22-1 Roppongi, Minato-ku, Tokyo 106, Japan
Tel: (+81-3)-3402-6231 ext. 2660-2663 Fax: (+81-3)-3402-4165

WorldAid'98   PALEXPO GENEVA
6-8 OCTOBER 1998

WorldAid is expanding fast, following its successful first global expo and conference on emergency relief in Geneva. It's launching a useful Web site, detailed report, massive directory and comprehensive CD-ROM guide, and has begun planning its next big event.

WorldAid'98 takes place Tuesday 6 October – Thursday 8 October 1998 at Palexpo Geneva, with a major expo of relief products and services, suppliers and agencies, and vital conference on the future of aid; procurement, standards, funding, safety, media and more.

Until then, WorldAid offers all those with a stake in disasters – agencies, companies, governments, military forces, foundations, consultants and the media – essential tools to keep up to date in the fast-moving world of emergencies and relief.

Copy this page and fax or mail it back to: WorldAid/ICVA, 13 rue Gauthier, CP 216,
1211 Geneva 21, Switzerland. Fax: +41(22)-738-9904

PLEASE COMPLETE IN BLOCK LETTERS. All prices include post and packing.

[   ] copy/copies of the 24-page *CROSSLINES Special Supplement* on all the issues tackled by WorldAid'96; available now: CHF 7.50 per copy.

[   ] copy/copies of the 160-page full-colour *WorldAid'96 Review and CD-ROM*, with a detailed guide to the WorldAid'96 expo and conference, full exhibitor listings; available now: CHF 22 per copy.

[   ] copy/copies of the *WorldAid'97 Directory*, listing thousands of suppliers and agencies involved in relief, from food and tools to drugs and vehicles; delivery on publication: CHF170 per copy.

[   ] copy/copies of the *WorldAid'97 Directory on CD-ROM*. Searchable by sector, name, location, etc., and clickable to reach web sites or send e-mail; delivery on publication: CHF170 per copy.

[   ] copy/copies of the *Reality of Aid 1996*, the 244-page ICVA/Eurostep guide to all aspects of aid spending, country profiles, tables and charts; available now: CHF 30 per copy.

[   ] copy/copies of the *Reality of Aid 1997*, the forthcoming edition with even more information. Get the facts at your fingertips; delivery on publication: CHF 30 per copy.

[   ] information on adding your agency, company, consultancy etc to the WorldAid Directory and CD-ROM global database: free.

[   ] *WorldAid'98.* Full information and registration pack as soon as it is published: free.

Card: AmEx/Visa/Diner's/MasterCard/Access/EuroCard   Expiry [/   ]

Number _____ TOTAL IN CHF _____

Name on card _____ Signature_____

Card billing address _____

Full name and job title _____

Delivery address (if different from billing address) _____

_____

Phone _____ Fax _____ E-mail _____

Organization type/sector (agency, company etc/ health, shelter etc) _____

_____ Web site _____

Did you or your organization attend or have a role at WorldAid'96?   Yes [   ]   No [   ]

Check out WorldAid at http://www.WorldAid.org

For information (no orders) e-mail WorldAid at icva_qva@gatekeeper.unicc.org

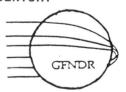

Professional and Volunteer Members Welcome

<u>MISSION</u>

Preparedness - Response - Development

Our members work together as a world-wide professional network of disaster researchers, response and recovery specialists, trainers, consultants, technical experts, and project managers.

We help disaster victims by improving communications and logistics, working to mitigate hazards, conducting community planning workshops, and by sponsoring emergency response teams.

We sponsor a school awards program that encourages students to study the effects of disasters and devise new strategies for effective mitigation and community preparedness.

As a prominent international professional association, our membership is composed of key leaders in the field of emergency management from around the world, as well as volunteers, business people, and students.

If you share our vision of commitment and service, we would welcome you as a member. Please write for a free brochure and copies of recent newsletters.

**DERA
P.O. Box 280795
Denver, CO 80228 USA**

Also, visit our World Wide Web Site at http://www.disasters.org/dera.html

Development in Practice

Development in Practice is a forum for practitioners, academics, and policy makers to exchange information and analysis concerning the social dimensions of development and emergency relief work. As a multi-disciplinary journal of policy and practice, *Development in Practice* reflects a wide range of institutional and cultural backgrounds and a variety of professional experience. *Development in Practice* is published by Oxfam UK & Ireland four times a year; all articles are independently refereed.

Editor Deborah Eade, Oxfam UK & Ireland
Reviews Editor Caroline Knowles, Oxfam UK & Ireland

Development in Practice Readers

A series of thematic selections of papers from past issues of *Development in Practice*. Titles for 1997 include *Development for Health* (March 1997, introduced by Eleanor Hill) and *Development and Patronage* (September 1997, introduced by Melakou Tegegn, El Taller). Titles already available are *Development and Social Diversity* (introduced by Mary B Anderson of the Collaborative for Development Action) and *Development in States of War* (introduced by Stephen Commins, UCLA and World Vision International).

Guidelines for Contributors are available from the Editor:

■ *Development in Practice*
c/o Oxfam
274 Banbury Road
Oxford, OX2 7DZ
UK

Subscriptions Concessionary rates are available for organisations from developing countries. For a free sample copy and details of subscription rates, please contact:

■ Carfax Publishing Company
PO Box 25
Abingdon OX14 3UE
UK
fax +44 (0)1235 401551

■ *In North America, contact:*
Carfax Publishing Company
875–81 Massachusetts Avenue
Cambridge MA 02139
USA
fax +1 617 354 6875

Send for a sample copy now

Now also available on the internet

Oxfam UK and Ireland is registered as a charity no. 202918 and is part of Oxfam International

◉ OXFAM

Mark Your Calendars <u>Now</u> for

NCCEM's 1997 Annual Conference & Exhibit

September 13-16, 1997

The Buttes Mountaintop Resort

Tempe, Arizona USA

- Disasters du jour;
- Unique solutions to common problems;
- New emergency management techniques;
- Innovative uses of technology;
- Ways to do more with less!

For additional meeting information call the *National Coordinating Council on Emergency Management* Headquarters at 001-703 538-1795

EARTHSCAN

NEW TITLES FROM EARTHSCAN

THE REALITY OF AID
An Independent Review of International Aid
1997-98

Edited by Tony German and Judith Randel of Development Initiatives for EUROSTEP and ICVA

'Indispensable..it gives you most of the hard facts you need to know about the major issues, plus a country-by-country breakdown of all the major aid donors' NEW INTERNATIONALIST

'should be on the shelf of any academic, student, NGO activist or politician with an interest in aid issues. It should also be required reading for donor agency officials' DEVELOPMENT AND CHANGE

The Reality of Aid, now in its fifth annual edition, provides a unique independent evaluation of the aid policies of the major donors and their effectiveness. A highly accessible and up-to-date reference, it is essential reading for all concerned with aid and development policy.

Contents: Introduction; Executive Summary; Aid and Poverty Eradication; The Uses of Aid (22 reports on how aid is spent by each OECD donor country and the EU); Perspectives from the South: Aid Reports from Eight Developing Countries; Aid Trends: An Overview; At a Glance Comparisons of the Outlook for Aid; Public Attitudes and Political Support; Development Education and Information; Progress on Measuring Aid to Poverty Eradication; Examples of Effective Aid; Gender; Environment; Relief Conflict and Emergencies; Aid Spent through NGOs; Glossary; Member agencies of ICVA and EUROSTEP.

£14.95 Paperback ISBN 1 85383 479 3 September 1997 224 pages (1996-97 edition currently available, also £14.95)

STRIKING A BALANCE
A Guide to Enhancing the Effectiveness of Non-Governmental Organisations in International Development
Alan Fowler

'an immensely useful "how to" tool for NGO leaders in development. One of the most powerful themes is the achievement potential of interaction between governments and NGOs. This book is an invaluable guide to both sides on how to achieve and what to expect from this bridging' JOHN CLARK, THE WORLD BANK

At a time of rapid global change, NGOs involved in international development are confronted with demands to simultaneously increase the scale of their impact, diversify their activities, respond to long-term humanitarian crises and improve their performance. *Striking a Balance* provides a practical guide to how NGDOs can respond better to these expectations.

£14.95 Paperback ISBN 1 85383 325 8 1997 320 pages

TO ORDER, CONTACT: **Earthscan Publications Ltd, 120 Pentonville Road, London N1 9JN**

Tel: (0)171 278 0433 Fax: (0)171 278 1142 Email: earthsales@earthscan.co.uk

Visit the Earthscan Web Site at http://earthscan.co.uk

DATE DUE

APR 0 9 1998		

Demco, Inc. 38-293